T0245321

ENERGY MANAGEMENT IN HOMES AND RESIDENTIAL MICROGRIDS

ENERGY MANAGEMENT IN HOMES AND RESIDENTIAL MICROGRIDS

SHORT-TERM SCHEDULING AND LONG-TERM PLANNING

REZA HEMMATI

Department of Electrical Engineering, Kermanshah University of Technology, Kermanshah, Iran

ELSEVIER

Elsevier
Radarweg 29, PO Box 211, 1000 AE Amsterdam, Netherlands
The Boulevard, Langford Lane, Kidlington, Oxford OX5 1GB, United Kingdom
50 Hampshire Street, 5th Floor, Cambridge, MA 02139, United States

Copyright © 2024 Elsevier Inc. All rights reserved.

No part of this publication may be reproduced or transmitted in any form or by any means, electronic or mechanical, including photocopying, recording, or any information storage and retrieval system, without permission in writing from the publisher. Details on how to seek permission, further information about the Publisher's permissions policies and our arrangements with organizations such as the Copyright Clearance Center and the Copyright Licensing Agency, can be found at our website: www.elsevier.com/permissions.

This book and the individual contributions contained in it are protected under copyright by the Publisher (other than as may be noted herein).

MATLAB® is a trademark of The MathWorks, Inc. and is used with permission. The MathWorks does not warrant the accuracy of the text or exercises in this book. This book's use or discussion of MATLAB® software or related products does not constitute endorsement or sponsorship by The MathWorks of a particular pedagogical approach or particular use of the MATLAB® software.

Notices
Knowledge and best practice in this field are constantly changing. As new research and experience broaden our understanding, changes in research methods, professional practices, or medical treatment may become necessary.

Practitioners and researchers must always rely on their own experience and knowledge in evaluating and using any information, methods, compounds, or experiments described herein. In using such information or methods they should be mindful of their own safety and the safety of others, including parties for whom they have a professional responsibility.

To the fullest extent of the law, neither the Publisher nor the authors, contributors, or editors, assume any liability for any injury and/or damage to persons or property as a matter of products liability, negligence or otherwise, or from any use or operation of any methods, products, instructions, or ideas contained in the material herein.

ISBN: 978-0-443-23728-7

For Information on all Elsevier publications
visit our website at https://www.elsevier.com/books-and-journals

Publisher: Megan R. Ball
Acquisitions Editor: Peter Adamson
Editorial Project Manager: Sara Valentino
Production Project Manager: Surya Narayanan Jayachandran
Cover Designer: Greg Harris

Typeset by MPS Limited, Chennai, India

Working together
to grow libraries in
developing countries

www.elsevier.com • www.bookaid.org

Contents

Preface

The widespread application of fossil fuels for electricity generation has created many issues such as environmental pollution, greenhouse gases, energy price hikes, and climate change. The countries aim to deal with fossil fuel issues in two ways: integration and development of renewable energies on the generation side and managing the users' energy consumption on the demand side. The home energy management system (HEMS) and residential microgrid energy management system (MEMS) are some efficient tools that facilitate renewable integration into the grid and at the same time, they can manage the users' load demand. This book presents detailed models and discussions for HEMS and MEMS, taking into account various practical situations and conditions for both short-term and long-term problems.

This book is primarily intended to be an informative resource on energy management in small energy systems such as homes and residential microgrids, suitable for students, engineers, and researchers wishing to specialize in the field of energy and electrical engineering. The book is also suitable for postgraduate courses at universities. An important part of the book is to develop the problems by mathematical modeling and numerical examples included in each chapter. The purpose is always to educate the readers and help them realize that many of the problems that will be faced in practice will need precise analysis, consideration, and some approximations.

The book is structured in seven chapters. In Chapter 1, the operation of traditional homes and moving toward modern homes is discussed. In Chapter 2, the HEMS in the homes integrated with renewable resources, loads, and energy storage systems is addressed. In Chapter 3, the impact of electric vehicles on energy management in homes and residential microgrids is discussed. In this chapter, the electric vehicle parking stations, charging stations, and emerging battery swapping stations are developed and modeled. In Chapter 4, the long-term capacity expansion planning in residential microgrids integrated with renewable resources, nonrenewable resources, and energy storage systems is talked about. In Chapter 5, the future of home energy management systems is discussed within the district energy paradigm. In Chapter 6, the integration of microgrids and smart homes into the electric distribution grids is addressed. In Chapter 7, a discussion on the dynamic models and control of renewable resources, storage devices, and AC−DC loads in small energy systems and microgrids is provided.

My colleagues in the educational community have also been instrumental in getting me started on this project, and I hope they find this book valuable and informative. No doubt, some errors will remain there, and I will be grateful if readers bring those errors to my attention.

Reza Hemmati

CHAPTER

1

Basic concepts of energy management in homes and residential microgrids

1.1 Introduction

In recent decades, the issues like barriers to fossil fuels, environmental pollution, greenhouse gasses, energy price, and climate change have motivated societies for moving toward the integration and development of renewable energies. Replacing the fossil fuels such as oil, gas and coal with renewable energy results in cutting the energy price, and

Energy Management in Homes and Residential Microgrids
DOI: https://doi.org/10.1016/B978-0-443-23728-7.00004-7

1

© 2024 Elsevier Inc. All rights reserved.

reducing environmental pollution and greenhouse gasses. Various renewable energies like wind turbines, solar photovoltaic systems, hydrogen storage reservoirs, hydropower plants, biomass generating systems and geothermal have been broadly investigated and built to deal with aforementioned issues. It is anticipated that renewable energy will produce about 60% of all electricity in the world. According to Eurostat data, in 2021, renewable energy represented 21.8% of the energy consumed in the European Union. Fig. 1.1 shows the share of renewables in the European Union energy in 2020 [1].

However, renewables also come with some issues like intermittency, uncertainty in production, and behavior and difficulty to store while fossil fuels can be easily stored and transported to the consumers. The current energy infrastructure and technology in the world are based on the use of fossil fuels and moving to renewables needs a significant change in the infrastructure and technology.

Apart from the integration of renewables into the electric generation systems, the other way to reduce the negative impacts of fossil fuels is to manage the energy consumed by loads and end users. In this regard, many concepts and models have been proposed to manage energy consumption on the demand side. The home energy management system (HEMS) [2] and microgrid energy management system (MEMS) [3] are some of the efficient ways to deal with issues of energy consumption on the demand side. A HEMS usually uses hardware and software to manage the energy consumption in a building which can be a residential, commercial, complex, etc. The optimal management of appliances in the home is the basic strategy in HEMS. However, many homes are integrated with renewables like rooftop solar panels and wind turbines and these energy resources can come into operation and change the objectives

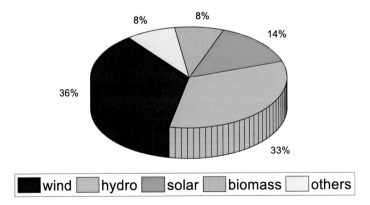

FIGURE 1.1   Electricity production by renewable energies in the European Union in 2020 taken from Eurostat data.

and models of HEMS. Some homes may also be benefited from a battery energy storage system to store electricity during some hours and usage at the next hours. Integration of renewables and storage technologies has changed the HEMS with regard to objectives, barriers, constraints and mathematical models [4].

1.2 Energy consumption in traditional homes

In traditional homes, the home only comprises loads and receives energy from the utility grid. Fig. 1.2 shows the typical traditional homes. All the homes appear as the load on the grid. In such systems, the HEMS can only manage the pattern of energy consumption by shifting loads from one hour to another hour. For instance, the received energy from the grid can be reduced during on-peak loading hours when the electricity is expensive and the loads can be supplied during off-peak loading hours when the electricity is often inexpensive. Such time-of-use energy consumption reduces the energy bill properly.

Fig. 1.3 shows a typical load pattern for a residential building and Fig. 1.4 presents a typical energy price in the residential area. It is seen that the energy price is high during on-peak hours like 13−14 and 18−22. The total cost of energy consumed by this building is calculated as Eq. (1.1), where, Ec shows electricity cost, Po_t indicates power, Pr_t is the electricity price and t is the index of time. The daily electricity cost for this building is 31.05 \$/day.

$$Ec = \sum_{t=1}^{24} (Po_t \times Pr_t) \tag{1.1}$$

The HEMS can shift energy from on-peak hours to off-peak hours in order to reduce bill costs. For instance, proper load shifting is shown in Fig. 1.5. Some loads are appropriately shifted over a 24-hour time period. It is seen that the loads are shifted from hours 12−15 and 18−22 to the initial hours of the day when there is low-priced electricity. The daily electricity cost of the building is now 26.2 \$/day which shows a 15.6% reduction compared to the first case. The annual value of bill reduction is more than 9500 \$/year which shows a significant cut in costs. The data of the proposed example are presented in detail in Table 1.1.

Regarding the load shifting, some points must be taken into account. All the appliances and electrical loads of the homes and residential microgrids cannot be shifted, interrupted, or curtailed [5]. As a result, a detailed look at the operation of loads is required [6]. Some loads in the homes are very important and can be neither shifted nor curtailed such as refrigerators. These loads may be assumed as constant or fixed loads. Some other loads can be shifted like laundry machines but when they

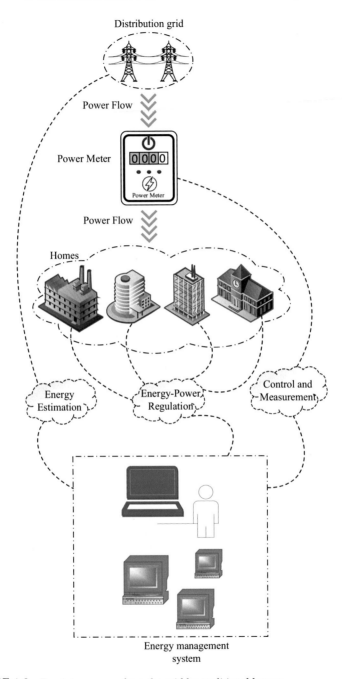

FIGURE 1.2   Receiving energy from the grid by traditional homes.

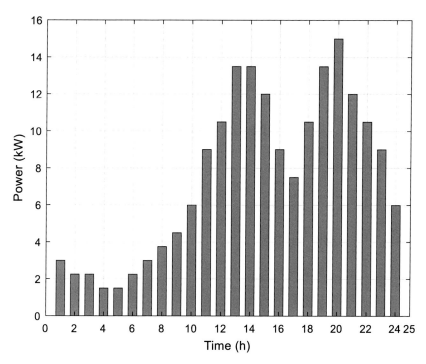

FIGURE 1.3   Typical load pattern for a residential building.

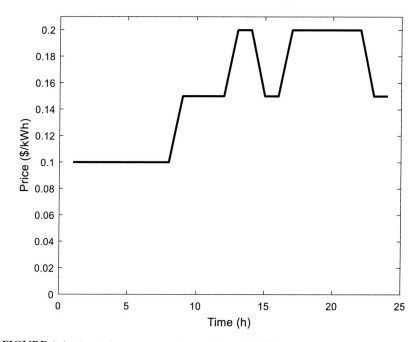

FIGURE 1.4   Typical energy price for a residential building.

Energy Management in Homes and Residential Microgrids

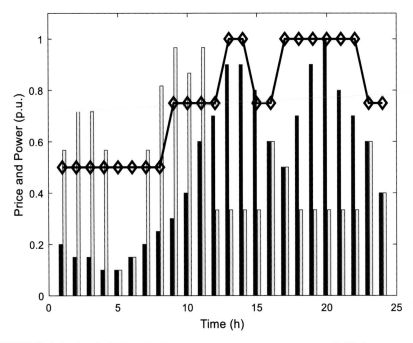

FIGURE 1.5 Load shifting by home energy management system. Solid line: energy price (per-unit); yellow bar: shifted load (per-unit); blue bar: original load (per-unit).

start working, they cannot be switched off until finishing their task. Hence, such loads are shiftable but not Interruptible and they need to be connected to the electricity for some consecutive hours and not separate hours. The other loads are noncritical loads which can be either curtailed or interrupted such as backyard lighting or unnecessary lighting systems. Some other loads can be interrupted and connected to the electricity in the next hours such as the devices that have the battery and need to be charged (e.g., electric vehicles). These rechargeable devices need specific energy for their battery (e.g., 50 kWh for electric vehicles) and such energy can be charged into their battery at various nonconsecutive hours. There is another group of loads that can be managed to consume less energy such as heating systems or air conditioning systems. Such loads are controlled and their energy is managed when required subject to the comfort and well-being of users. With respect to the aforementioned topics, the loads are categorized as listed as follows:

- Fixed load;
- Interruptible load;
- Curtailable load;
- Managed loads;
- Shiftable loads.

TABLE 1.1 Data of the proposed example.

Time (h)	Original load (kW)	Shifted load (kW)	Energy price ($/kWh)
1	3	8.5	0.1
2	2.25	10.75	0.1
3	2.25	10.75	0.1
4	1.5	8.5	0.1
5	1.5	1.5	0.1
6	2.25	2.25	0.1
7	3	8.5	0.1
8	3.75	12.25	0.1
9	4.5	14.5	0.15
10	6	13	0.15
11	9	14.5	0.15
12	10.5	5	0.15
13	13.5	5	0.2
14	13.5	5	0.2
15	12	5	0.15
16	9	9	0.15
17	7.5	7.5	0.2
18	10.5	5	0.2
19	13.5	5	0.2
20	15	5	0.2
21	12	5	0.2
22	10.5	5	0.2
23	9	9	0.15
24	6	6	0.15

From another point of view, the loads may be classified into two groups critical and noncritical loads. The critical loads are very important and need to be supplied under any situation such as outages and events. The noncritical loads have no priority and may be curtailed when the utility grid faces events and outages. The HEMS can prioritize the loads and shed the noncritical loads when there is a shortage in energy supply.

1.3 Energy consumption in modern homes

In modern homes, the home is integrated with renewable resources and even energy storage systems. Fig. 1.6 shows a typical modern home integrated with renewables and energy storage. In this paradigm, the home can receive power from the external grid or send power to the grid. As a result, the exchanged power with the grid needs to be scheduled. The solar and wind energies supply the building and the battery storage unit is used to shift energy based on economic or technical dictates. In such homes, the HEMS aims to do the following items [7]:

- Changing the pattern of energy consumption by shifting loads.
- Designing an optimal charging scheduling for the energy storage system.
- Harvesting maximum power from renewable resources.
- Designing an optimal operating pattern for nonrenewable resources such as diesel generators.

The HEMS may follow a variety range of objectives like bill reduction, off-grid operation, acting as backup for the grid, critical load supplying, enhancing the reliability of loads and improving the resilience of the home after events. In conventional homes, the bill reduction is done by managing the received power from the grid and shifting loads over the day hours, but here the bill reduction can be done by managing the received power from the grid as well as exporting and selling energy to the external grid. In the on-peak hours when the grid needs power, the HEMS can send excess power to the grid for making a profit. In order to achieve the minimum bill cost, the problem needs to be defined as optimization programming and solved by available software like GAMS. In such problems, the charging schedule is optimized for batteries and the receiving-sending power to the external grid is also managed and optimized in order to attain maximum profit or minimum cost.

The off-grid operation of such homes is also possible unlike the traditional homes, where the off-grid operation is not feasible [8,9]. Fig. 1.7 shows the off-grid operation of the system. In the off-grid state, the exchanged power with the grid is zero and all the consumed energy by the building must be generated by its own resources. In such systems, energy storage plays a major role. As seen in Fig. 1.7, the only element of the home that can receive energy is the energy storage unit. The excess energy needs to be stored in the storage unit and then restored for the use of loads. The most important point is the mismatch between the load power profile and solar-wind power profiles. Generally, the solar and wind energy profiles do not match the load energy pattern and there will be a mismatch of energy between generation and load demand. Such a mismatch can be efficiently addressed by an energy storage system.

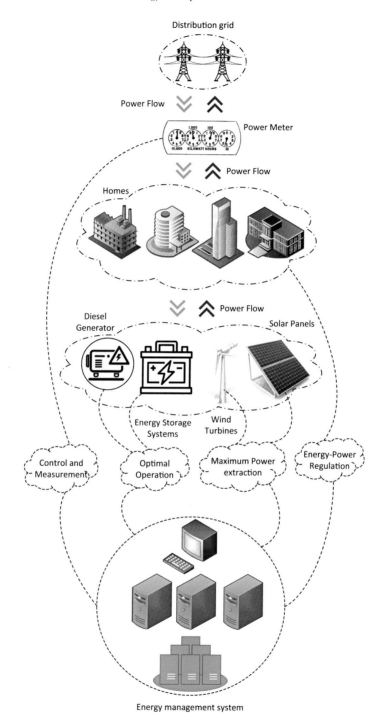

FIGURE 1.6 Homes equipped with wind, solar and battery systems.

Energy Management in Homes and Residential Microgrids

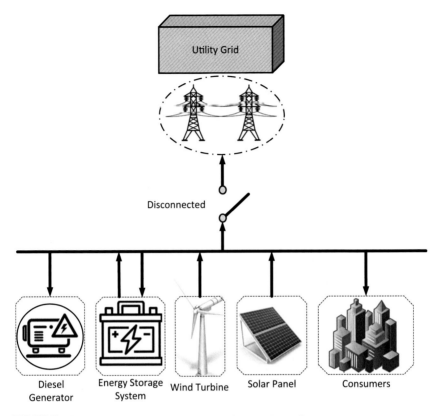

FIGURE 1.7   Off-grid home integrated with renewables and storage system.

A typical solar energy profile together with a typical load profile is shown in Fig. 1.8 and the mismatch between those two profiles is presented in Fig. 1.9. It is clear that there is a significant mismatch especially when the load is on the peak hours like hours 18−24. In other words, when the home is supplying the highest loads, the solar energy becomes zero. Such mismatch needs to be dealt with in the HEMS. At some hours the solar power is more than the load power and there is a negative mismatch, and such surplus of energy must be stored by the energy storage system.

Similar to solar energy and load, there is a mismatch between wind energy and load as well. Figs. 1.10 and 1.11 demonstrate wind energy and its mismatch with load demand. It is clear that the mismatch of energy in this situation is more volatile compared to the previous condition. It is because of wind power alterations. The negative area in the mismatch shows the time periods when wind energy is more than load demand. The data of this example are presented in detail in Table 1.2.

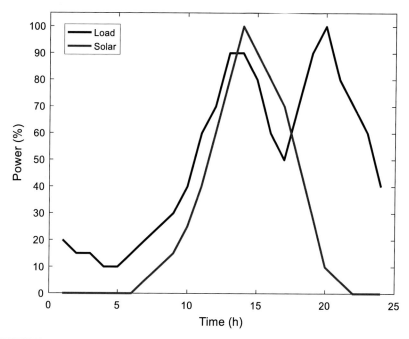

FIGURE 1.8   Load and solar energy profiles.

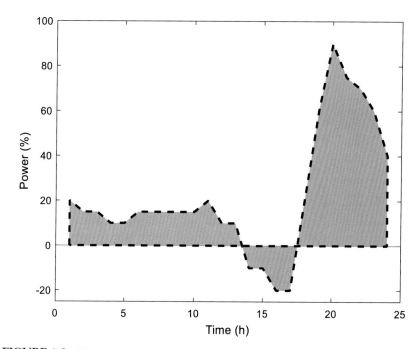

FIGURE 1.9   Mismatch between load and solar energy profiles.

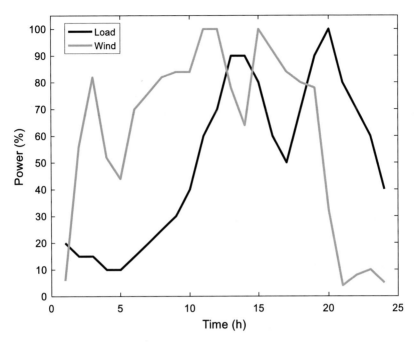

FIGURE 1.10   Load and wind energy profiles.

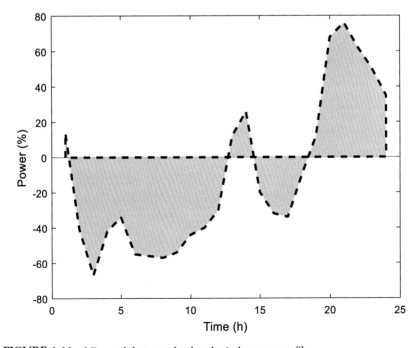

FIGURE 1.11   Mismatch between load and wind energy profiles.

TABLE 1.2 Wind and Solar energy profiles.

Time (h)	Solar power (%)	Wind power (%)
1	0	6
2	0	56
3	0	82
4	0	52
5	0	44
6	0	70
7	5	76
8	10	82
9	15	84
10	25	84
11	40	100
12	60	100
13	80	78
14	100	64
15	90	100
16	80	92
17	70	84
18	50	80
19	30	78
20	10	32
21	5	4
22	0	8
23	0	10
24	0	5

If both the introduced wind and solar systems are integrated into the home, the mismatch of energy will be as shown by Fig. 1.12. In this case, the renewable energy volatility is reduced because wind and solar somehow support and complement each other. A proper charging–discharging process of an energy storage system can deal with such energy mismatch as shown in Fig. 1.13. The excess energy is stored when the generation is more than the load demand, then the energy is discharged and injected into the home when the load is greater than the generation.

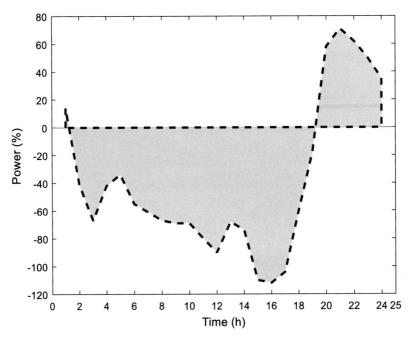

FIGURE 1.12   Mismatch between load and wind–solar energy profiles.

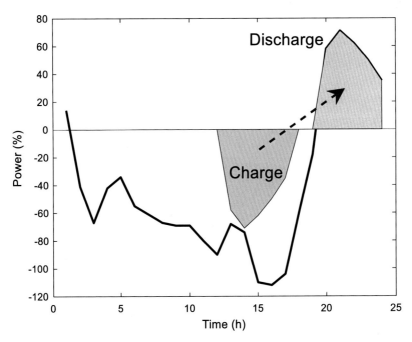

FIGURE 1.13   Operation of energy storage to deal with energy mismatch.

1.4 Wind and solar system modeling

Fig. 1.14 shows the output power of the wind turbine versus wind speed. At low wind speeds, the output power is zero. Then the output power increases together with wind speed. In the rated wind speed, the output power will be fixed on the rated power and it will be controlled by the pitch angle of the wind turbine. Eventually, if the wind speed goes beyond the cut-out speed, the wind turbine will be switched off to avoid damage. The mathematical model of such an operation is given by Eq. (1.2) [10].

$$P_W = \begin{cases} 0 & W_s < W_s^{ci} \\ f(W_s) & W_s^{ci} \le W_s \le W_s^{r} \\ P_W^{max} & W_s^{r} \le W_s \le W_s^{co} \\ 0 & W_s \ge W_s^{co} \end{cases} \qquad (1.2)$$

where P_W is the output power of wind turbine (kW), W_s is the wind speed (m/s), W_s^{ci} is the cut-in, peed of wind turbine (m/s), W_s^{r} is the rate speed of wind turbine (m/s), W_s^{co} is the cut-out speed of wind turbine (m/s), and $f(W_s)$ is the function showing the relationship between wind speed and output power.

In the mechanical section of a wind turbine, the power is expressed as Eq. (1.3). The parameters of this relationship are expressed through Eqs. (1.4)–(1.6) [11].

$$Pm = 0.5\rho CpV_{\omega}^3 \qquad (1.3)$$

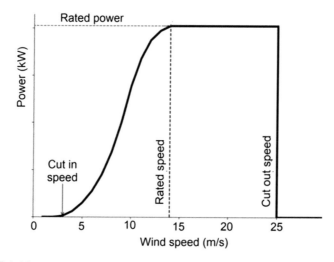

FIGURE 1.14   Output power of wind turbine versus wind speed.

$$Cp(\lambda, \theta) = 0.73 \left[\frac{151}{\lambda_i} - 0.58\theta - 0.002\theta^{2.14} - 13.2 \right] \exp\left(-\frac{18.4}{\lambda_i} \right) \qquad (1.4)$$

$$\frac{1}{\lambda_i} = \frac{1}{\lambda + 0.08\theta} - \frac{0.035}{\theta^3 + 1} \qquad (1.5)$$

$$\lambda = \frac{R * \omega_r}{V_\omega} \qquad (1.6)$$

where the parameters are defined as follows:

A	swept area (m²),
C_p	power coefficient,
P_m	turbine output power (W),
V_ω	wind speed (m/s),
ρ	air density (kg/m²),
ω_r	turbine rotational speed (m/s),
θ	pitch angle (deg),
λ	tip-speed ratio.

The produced current by the solar photovoltaic system is modeled by Eq. (1.7). The diode current is specified by Eq. (1.8) and the solar cell output power is modeled by Eq. (1.9) [11].

$$Ipv = Rk_c \left[I_{sq} + a(Tc - Tc_{ref}) \right] - ID - Ish \qquad (1.7)$$

$$ID = \left[K * T^{(3+\gamma/2)} \left[\exp\left(-\frac{E_g}{nK_BT} \right) \right] \right] \times \left[\exp\left(\frac{q(Vpv + (Ipv * Rs))}{nK_BT} \right) \right] \qquad (1.8)$$

$$Ppv = Vpv * Ipv \qquad (1.9)$$

where the parameters are defined as follows:

a	temperature coefficient of short circuit current (A/C⁰),
E_g	bandgap energy,
ID	current of diode (A),
Ipv	current of solar-cell (A),
K_C	concentration ratio,

(Continued)

(Continued)

K_B	Boltzmann's constant (jk^{-1}),
n	diode ideality factor,
P_{PV}	power of solar-cell (W),
q	electron charge (C),
R	solar radiation (kW/m^2),
R_S	cell series resistance (Ohm),
T	absolute temperature (K),
T_{Cref}	reference temperature (C^0),
Vpv	voltage of solar-cell (V).

1.5 Uncertainty of loads and renewable energies

In the previous sections, the operations of renewable energies like wind and solar systems were discussed but, in actual systems, the output power produced by those resources is not constant and changes together with wind speed or solar irradiation and temperature. As a result, there is a parametric uncertainty in the output power of wind and solar systems. Solar energy is technically zero during the night and it increases from morning to afternoon and then reduced in the evening. Such behavior is shown in Fig. 1.15A, where a typical output power for a solar photovoltaic system is presented. However, as it was stated, the power is not constant and comprises volatilities and uncertainty. The uncertainty can be defined as a set of scenarios of performance. For instance, the nominal power at each hour can be added by a random number to generate a scenario of performance that shows a random operation of the system. In this regard, the nominal curve shown in Fig. 1.15A which is the nominal power of the solar system is added by a normal distribution with a standard deviation of 2. This point is shown by Eq. (1.10), where the $P_{so}^{t,s}$ shows solar power under each scenario at each hour, Pn_{so}^t is the nominal power of the solar system at each hour, $Randn$ generates random values from a normal distribution with mean 0 and standard deviation 1, and k shows the desired standard deviation. Fig. 1.15 shows the output power with 1, 3, 5 and 9 scenarios. Together with increasing the number of scenarios, the wider area of uncertainty is covered and the problem is close to the realistic operation of the system. On the other hand, increasing the number of scenarios results in a complicated mathematical model and increases the complexity of the problem and needs more time for simulation by computer.

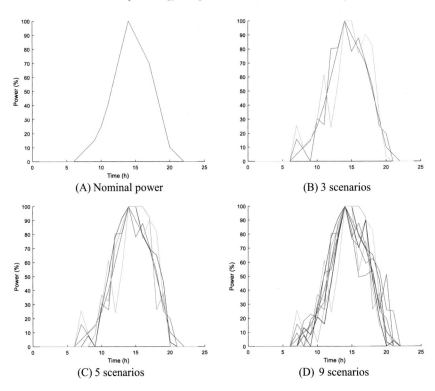

FIGURE 1.15   Output power of the solar system with different scenarios of performance.

$$P_{so}^{t,s} = Pn_{so}^{t} + k \times Randn$$
$$\forall s \in S \tag{1.10}$$
$$\forall t \in T$$

In order to evaluate the impacts of uncertainty on the problem, a home with solar energy and loads is modeled as shown in Fig. 1.16. The home is connected to the grid and can exchange energy with the external grid as expressed by Eq. (1.11).

$$P_{g}^{s,t} = P_{pv}^{s,t} - P_{l}^{t}$$
$$\forall t = [1:24] \tag{1.11}$$
$$\forall s = [1:9]$$

The parameters are defined as follows: $P_{g}^{s,t}$ is the exchanged power with grid (kW), $P_{pv}^{s,t}$ is the produced power by the solar photovoltaic system (kW), P_{l}^{t} is the consumed power by load (kW), t,s is the index of hours and index of scenarios.

The rated power of the solar system is equal to 15 kW and the peak-load power is also 15 kW. The solar energy profiles under nine scenarios

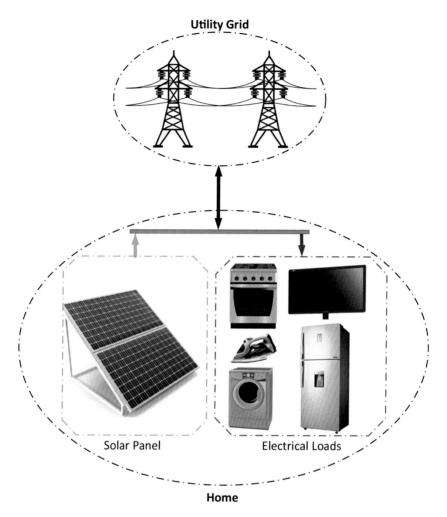

FIGURE 1.16 Home with solar energy and loads connected to the grid.

of performance are listed in Table 1.3. The load energy profile is also taken from Table 1.1. Figs. 1.17 and 1.18 show the exchanged power with the grid at hours 12 and 19 under all nine scenarios of performance, respectively. The exchanged power with the grid varies from one scenario of performance to another one. At hour 12, the solar power is high and the home can inject power into the grid under some scenarios such as 2 and 4. However, solar energy alteration changes power significantly. For instance, the home injects power to the grid under scenario 2 but takes an enormous power under scenario 3 and then injects power to the grid again under scenario 4. Such uncertainty in behavior makes significant impacts on HEMS

TABLE 1.3 Solar energy profile under all nine scenarios of performance.

Scenario number									Hours
1	2	3	4	5	6	7	8	9	
0	0	0	0	0	0	0	0	0	1
0	0	0	0	0	0	0	0	0	2
0	0	0	0	0	0	0	0	0	3
0	0	0	0	0	0	0	0	0	4
0	0	0	0	0	0	0	0	0	5
0	0	0	0	0	0	0	0	0	6
5	15.85	25.64	0	0	0.81	0	0	9.68	7
10	8.45	9.37	7.75	0	0	13.28	18.09	0	8
15	0	14.11	0	8.41	17.41	8.25	22.82	19.25	9
25	30.48	29.7	27.78	36.81	2.76	20.82	20.33	24.01	10
40	26.16	61.73	51.47	32.39	27.25	15.94	52.39	31.79	11
60	80.26	24.15	77.69	70.41	43.96	46.33	54.75	61.02	12
80	80.78	54.85	91.29	85.1	82.53	77.1	78.27	87.69	13
100	100	100	100	100	100	100	100	100	14
90	78.19	100	100	100	100	87.13	81.9	76.59	15
80	87.85	76.93	79.15	100	53.66	70.38	49.36	71.42	16
70	70.97	90.31	69.42	92.01	75.2	89.93	50.67	59.19	17
50	53.24	82.55	65.31	26.24	40.19	47.08	54.36	62.72	18
30	25.2	32.85	46.31	37.95	13.78	18.36	26.76	53.59	19
10	19.65	3.17	2.94	0	11.55	0	51.27	25.84	20
5	0	0	0	0	0	3.87	0	25.12	21
0	0	0	0	0	0	0	0	0	22
0	0	0	0	0	0	0	0	0	23
0	0	0	0	0	0	0	0	0	24

and it is very difficult to make a contract with the upstream grid for exchanging power. At hour 12, since the solar power is close to zero, the home cannot inject any power into the grid and only receives power under all scenarios of performance. However, the taken power is very volatile. Table 1.4 presents the exchanged power with the grid under all scenarios of performance. The impacts of solar energy alteration on the exchanged

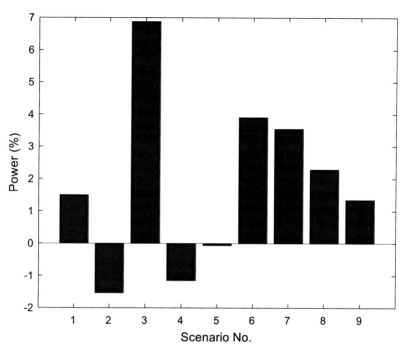

FIGURE 1.17 Exchanged power with the grid at hours 12 under all nine scenarios.

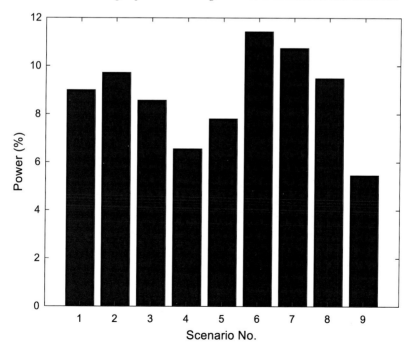

FIGURE 1.18 Exchanged power with the grid at hours 19 under all nine scenarios.

TABLE 1.4   Exchanged power with grid under all nine scenarios of performance.

Scenario number									Hours
1	2	3	4	5	6	7	8	9	
3	3	3	3	3	3	3	3	3	1
2.25	2.25	2.25	2.25	2.25	2.25	2.25	2.25	2.25	2
2.25	2.25	2.25	2.25	2.25	2.25	2.25	2.25	2.25	3
1.5	1.5	1.5	1.5	1.5	1.5	1.5	1.5	1.5	4
1.5	1.5	1.5	1.5	1.5	1.5	1.5	1.5	1.5	5
2.25	2.25	2.25	2.25	2.25	2.25	2.25	2.25	2.25	6
2.25	0.62	-0.85	3	3	2.88	3	3	1.55	7
2.25	2.48	2.34	2.59	3.75	3.75	1.76	1.04	3.75	8
2.25	4.5	2.38	4.5	3.24	1.89	3.26	1.08	1.61	9
2.25	1.43	1.55	1.83	0.48	5.59	2.88	2.95	2.4	10
3	5.08	-0.26	1.28	4.14	4.91	6.61	1.14	4.23	11
1.5	-1.54	6.88	-1.15	-0.06	3.91	3.55	2.29	1.35	12
1.5	1.38	5.27	-0.19	0.73	1.12	1.93	1.76	0.35	13
-1.5	-1.5	-1.5	-1.5	-1.5	-1.5	-1.5	-1.5	-1.5	14
-1.5	0.27	-3	-3	-3	-3	-1.07	-0.29	0.51	15
-3	-4.18	-2.54	-2.87	-6	0.95	-1.56	1.6	-1.71	16
-3	-3.15	-6.05	-2.91	-6.3	-3.78	-5.99	-0.1	-1.38	17
3	2.51	-1.88	0.7	6.56	4.47	3.44	2.35	1.09	18
9	9.72	8.57	6.55	7.81	11.43	10.75	9.49	5.46	19
13.5	12.05	14.52	14.56	15	13.27	15	7.31	11.12	20
11.25	12	12	12	12	12	11.42	12	8.23	21
10.5	10.5	10.5	10.5	10.5	10.5	10.5	10.5	10.5	22
9	9	9	9	9	9	9	9	9	23
6	6	6	6	6	6	6	6	6	24

powers can be easily seen. At hours 1–6 and 22–24 when the solar power is zero, the traded power with the external grid is constant under all scenarios of performance.

Wind energy has similar impacts on the HEMS. In order to show the impacts of wind power uncertainty on the problem, the previous example is re-formulated with wind energy as modeled by Eq. (1.12), where

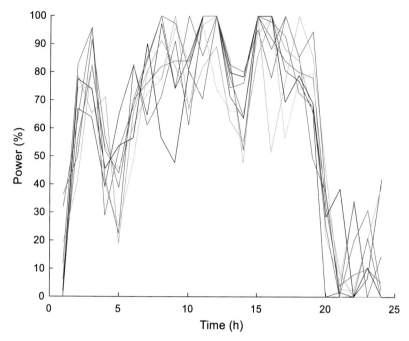

FIGURE 1.19   Output power of wind turbine under nine scenarios of performance.

the $P_w^{s,t}$ shows the wind power under each scenario at any hour. The rated power of the wind system is 15 kW.

$$P_g^{s,t} = P_w^{s,t} - P_l^t$$
$$\forall t = [1{:}24]$$
$$\forall s = [1{:}9]$$

(1.12)

The wind energy profiles under nine scenarios of performance are shown in Fig. 1.19. Fig. 1.20 demonstrates the exchanged power with the grid at hour 1 under all nine scenarios. Under some scenarios, the power is sent to the grid, and under some other scenarios, the power is taken from the grid while the taken and sent powers vary from one scenario to another one. Such uncertainty entirely changes HEMS and must be dealt with properly. Table 1.5 presents the traded power with the external grid under all scenarios of performance.

The solar, wind, and load energies all can be affected by uncertainty. In such a situation, the HEMS needs to use uncertainty management systems like stochastic programming or robust programming to deal with uncertainty in the problem. Fig. 1.21 shows an outline of such a problem. In this case, the level of uncertainty in the problem increases and it would need more than nine scenarios to simulate the uncertainty. Fig. 1.22 depicts the modeling of the wind energy profile with 10 and

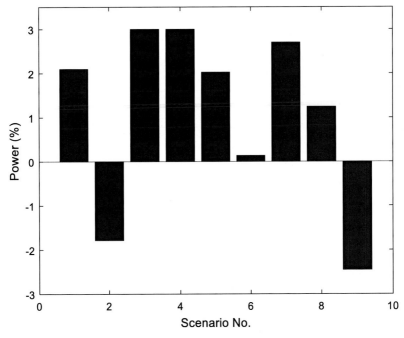

FIGURE 1.20   Exchanged power with the grid at hour 1 under all scenarios of performance.

TABLE 1.5   Exchanged power with grid under all scenarios of performance.

				Scenario number					Hours
1	**2**	**3**	**4**	**5**	**6**	**7**	**8**	**9**	
2.1	−1.78	3	3	2.02	0.14	2.7	1.25	−2.45	1
−6.15	−7.29	−8.91	−7.79	−9.55	−3.87	−9.38	−10.18	−5.22	2
−10.05	−12.14	−8.9	−7.34	−7.55	−10.16	−8.84	−12.1	−11.48	3
−6.3	−2.83	−4.46	−4.37	−9.17	−6.49	−5.33	−4.82	−7.02	4
−5.1	−6.38	−2.26	−8.3	−1.33	−5.31	−6.56	−1.87	−4.32	5
−8.25	−7.9	−4.95	−10.1	−7.62	−6.81	−6.22	−10.16	−8.24	6
−8.4	−9.64	−9.02	−6.7	−9.05	−7.89	−10.52	−6.17	−9.79	7
−8.55	−11.25	−7.8	−10.84	−9.95	−7.86	−4.76	−6.94	−11.25	8
−8.1	−6.66	−9.87	−6.62	−7.82	−10.5	−2.65	−9.15	−10.09	9
−6.6	−9	−6.47	−6.84	−4.16	−4	−6.24	−3.17	−6.05	10
−6	−3.88	−5.53	−6	−3.26	−5.54	−6	−6	−1.56	11
−4.5	−4.5	−0.63	−4.5	−2.88	−4.5	−4.5	−4.5	−4.5	12
1.8	1.11	3.54	2.8	4.09	0.85	1.53	2.33	1.21	13

(Continued)

TABLE 1.5   (Continued)

Scenario number									Hours
1	2	3	4	5	6	7	8	9	
3.9	1.5	3.67	3.97	5.2	6.35	1.75	2.08	5.68	14
−3	−2.27	−2.04	−3	−0.86	−3	−3	−3	−3	15
−4.8	−2.69	−6	−6	−6	1.28	−6	−4.24	−5.68	16
−5.1	−7.5	−0.98	−4.58	−5.71	−3.99	−2.92	−7.5	−6.42	17
−1.5	−2.32	−1.09	−1.03	−2.14	−4.5	−1.33	−2.33	−0.41	18
1.8	6.11	1.54	3.56	3.69	0.68	3.15	−0.68	3.34	19
10.2	9.48	11.83	15	9.58	8.42	10.73	9.61	11.25	20
11.4	12	11.15	11.75	10.67	11.61	6.23	11.83	12	21
9.3	7.52	5.56	10.5	10.5	10.16	10.5	5.37	10.48	22
7.5	4.38	9	8.01	9	5.93	7.42	9	5.86	23
5.25	5.66	6	−0.3	6	0.04	5.97	3.83	6	24

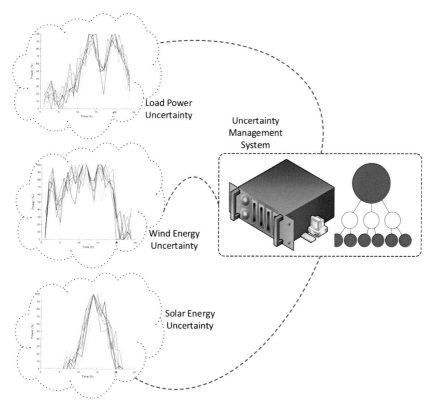

FIGURE 1.21   Problem with uncertainty in load, solar and wind energies.

100 scenarios. It is clear that when the number of scenarios increases, the uncertainty is covered better and the model behaves similarly to the realistic operation of the wind energy system.

1.6 Scenario generation and reduction techniques

Figs. 1.23 and 1.24 display the technique of scenario generation based on the following steps [12]:

Step 1: The distribution function of every uncertain parameter is modeled by a continuous function and it is essential to convert this model to a discrete function. This would create errors in the model that needs to be considered. The discrete function needs to be described by appropriate occurrence probabilities.

Step 2: A cumulative distribution function is also calculated for every uncertain parameter.

Step 3: The scenario generation technique is continued based on the roulette wheel mechanism. A roulette wheel is made by the relative fitness (ratio of individual fitness and total fitness) of each uncertain parameter. It is modeled as a pie chart where the area taken by each

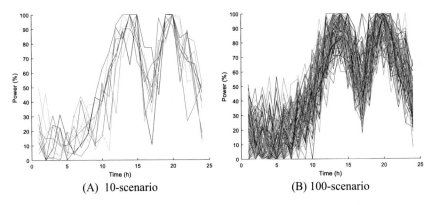

(A) 10-scenario   (B) 100-scenario

FIGURE 1.22   Wind energy profile modeled by 10 and 100 scenarios.

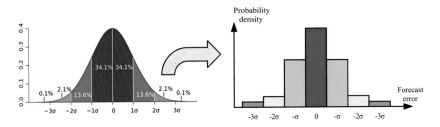

FIGURE 1.23   Discretization of a typical probability distribution function.

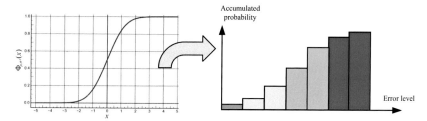

FIGURE 1.24   Accumulated normalized probability function.

uncertain parameter on the roulette wheel is proportional to its relative fitness.

Step 4: The parameters of each scenario are updated to proper new ones and new values are computed and gathered.

Step 5: The total probability of each scenario is calculated. The total probability is equal to the multiplication of all the computed probabilities of the parameters of each scenario.

These steps should be repeated until the desired number of scenarios is achieved.

Since the generated scenarios may be numerous, it is required to carry out scenario reduction techniques to reduce the number of scenarios to an acceptable level. The backward scenario reduction technique is one of the efficient methods and it is described here.

Step 1: Two sets of S and D_S are formed as the sets of acceptable and unacceptable scenarios. The D_S is initially empty. The distance of all couples of scenarios is calculated as follows:

$$DT_{s,s'} = DT\left(\xi_s, \xi_{s'}\right) = \sqrt{\sum_{i=1}^{d} \left(x_i^s - x_i^{s'}\right)} \quad s, s' \in S$$

where $DT_{s,s'}$ is the distance of a couple of scenarios, ξ_s is the set of scenarios, and s is the index of scenarios.

Step 2: The minimum distance of scenario pairs is calculated as follows:

$$DT_{k,r} = \min DT_{k,s'} \quad k, s' \in S; s \neq k$$

Step 3: $PD_{k,r}$ and PD_d are characterized as follows:
$PD_{k,r} = \rho(k) \times DT_{k,r} \quad k \in S$ where $\rho(s)$ is the probability of a scenario

Step 4: New sets of scenarios are obtained based on the following equation:
$PD_d = \min PD_k \quad k \in S$

Step 5: The needed reduction is achieved by repeating the previous steps.

Stochastic programming is based on the scenario generation and scenario reduction techniques [12]. The stochastic program with CVaR evaluated in a low-probability tail can provide better cost-risk trade-offs than the robust formulation with less conservative preferences [13].

1.7 Outages and events

The uncertainty shows the unpredictable behavior of energy resources or loads, but the outage indicates the events and situations that would occur because of natural disasters, terrorism, physical attacks, and cyber-attacks which lead to widespread outages and blackouts. Renewable energies are broadly affected by natural phenomena like storms and rain and their outage is likely to happen. As a result, considering renewable uncertainty is not enough and it is necessary to take into account their unavailability in the problems [14,15].

Fig. 1.25 shows an outage of wind energy for two consecutive hours while the outages can occur at nonconsecutive hours like in Fig. 1.26. The outage can also be associated with uncertainty and scenarios of performance as demonstrated by Fig. 1.27.

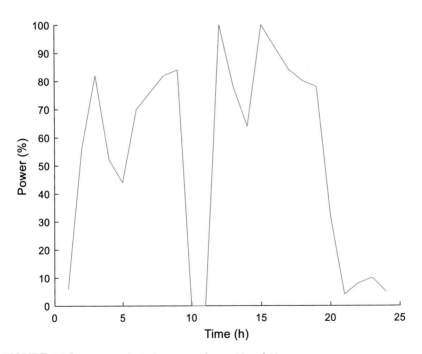

FIGURE 1.25   Outage of wind energy at hours 10 and 11.

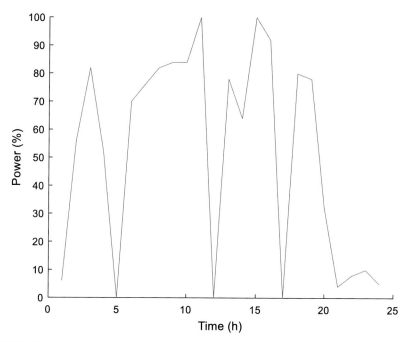

FIGURE 1.26 Outage of wind energy at hours 5, 12, and 17.

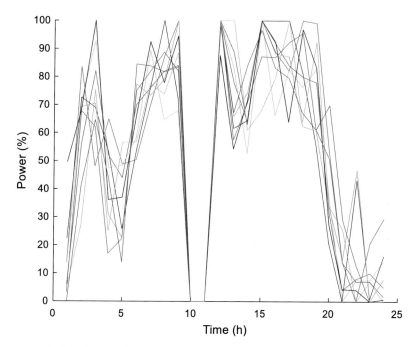

FIGURE 1.27 Outage of wind energy at hours 10 and 11 together with uncertainty.

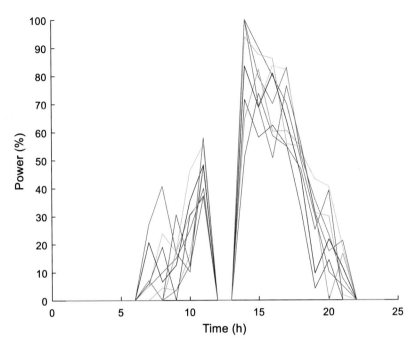

FIGURE 1.28   Outage of solar energy at hours 12 and 13 together with uncertainty.

The solar energy and loads may also face outage. Fig. 1.28 indicates the outage of solar energy at hours 12 and 13 together with uncertainty. The outage of solar energy is more likely compared to wind energy because of rain, clouds and shadows that can turn the solar system off. The loads of the grid may also be exposed to the events and switched off at some hours as shown in Fig. 1.29.

1.8 Advanced topics and the future of home energy

The resilience of a single carrier energy system based on electricity is not adequate following events and outages of the distribution grid. As a result, the multicarrier energy systems and energy hubs would form the future energy systems in the microgrids and smart-homes [7]. In multicarrier energy systems, natural gas and electricity are the main input energies to the system and the output carriers of energy are often cooling, heating, and electricity. In multicarrier energy systems, every form of energy can be converted to another form by proper energy converters such as combined heat and power (CHP) unit, micro-gas boiler, absorption-electrical chillers, power-to-gas (P2G) unit and fuel-cell systems [16]. The common energy storage technologies such as electrical

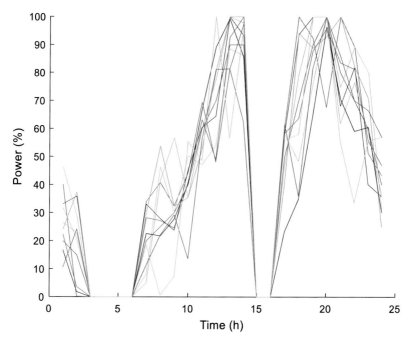

FIGURE 1.29 Outage of load at hours 3—6 and 15—16 together with uncertainty.

storage system, heating storage unit, cooling storage devices and hydrogen storage reservoirs are broadly integrated into multicarrier energy systems and energy hubs [17]. They operate various forms of energy simultaneously including electric power, heating, cooling, natural gas, solar, wind, water, hydrogen, and carbon dioxide. Fig. 1.30 depicts a typical multicarrier energy system with electricity and natural gas as input energies and electricity, heating and cooling as output energies. The storage devices are equipped for storing gas, electricity, heating and cooling energies. Each energy carrier is converted to other forms of energy by proper energy converters. Once one of the resources or energy carriers is faced with an outage, the other resources take over the responsibility for supplying the loads. As a result, the resilience of the system is significantly increased compared to the traditional single-carrier energy systems [3].

Table 1.6 lists the typical configurations of multicarrier energy systems. It is seen that electricity, heating, natural gas, cooling, and hydrogen are the common energy types in multicarrier energy systems. Electrical, heating, cooling, and gas storages are the regular energy storage technologies in such systems. Various forms of energy converters are used but heating to electricity, electricity to heating, heating to cooling, electricity to cooling, gas to electricity and electricity to gas are the

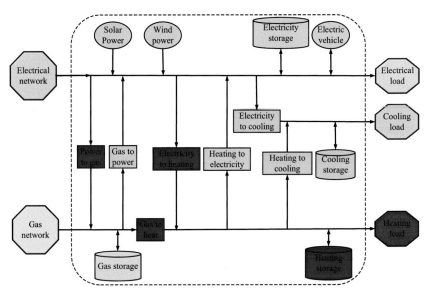

FIGURE 1.30   Typical multicarrier energy system.

TABLE 1.6   Typical configurations of multicarrier energy systems [3].

Reference number	Input energy carrier	Energy storage technology	Energy converter	Loading type
[18]	Electricity Heating Biogas	Biogas storage Electrical storage	Combined heat and power Boiler Furnace Digester	Electrical Heating Biogas
[19]	Electricity Heating Natural gas Cooling Carbon dioxide Hydrogen	Electrical storage Heating storage Cooling storage Gas storage	Combined heat and power Boiler Absorption chiller Electric chiller Power to gas	Electrical Heating Cooling
[20]	Electricity Heating Natural gas Hydrogen	Gas storage	Combined heat and power Boiler Power to gas	Electrical Heating Gas
[21]	Electricity Heating NG Cooling	Electrical storage Heating storage Gas storage	Combined heat and power Boiler Absorption chiller Electric chiller	Electrical Heating Cooling

(*Continued*)

TABLE 1.6 (Continued)

Reference number	Input energy carrier	Energy storage technology	Energy converter	Loading type
[22]	Electricity Heating NG Cooling	Electrical storage Heating storage	Combined heat and power Boiler Absorption chiller Electric chiller	Electrical Heating Cooling
[23]	Electricity Heating Natural gas Hydrogen	Electrical storage Heating storage Hydrogen storage	Combined heat and power Boiler Power to gas Fuel cell	Electrical Heating Hydrogen
[24]	Electricity Heating Natural gas Hydrogen	Electrical storage Heating storage Gas storage	Combined heat and power Boiler Power to gas	Electrical Heating Gas

common types of energy converters. The output energies are often electricity, heating and cooling [3].

The purpose of developing multicarrier energy systems is to minimize the operating cost of resources as well as the investment cost of new resources. The operating cost includes the costs related to different parts of the system such as electrical components, natural gas devices, energy converters, boilers, energy storage systems, combined heat-power units and maintenance cost of renewable resources. As well, some extra costs such as environmental pollution (or environmental protection) tax, load curtailment penalty cost and resilience cost may be included in the operating cost. The multicarrier energy systems may be owned by private stockholders and be operated to maximize the profit of owners, where the output energies like Methane, carbon, electricity, heating, and cooling are traded for making a profit. Environmental pollution reduction is one of the advantages achieved by multicarrier energy systems. The carbon tax policy, environmental pollution taxation, and environmental protection tax are the common indexes for consideration of environmental pollution in the mathematical model [3].

1.9 Conclusions

Together with increasing the issues and barriers of fossil fuels like environmental pollution, greenhouse gasses, energy price, and climate change, renewable energies are developed to deal with the concerns of electricity generation. Renewables come with some problems like intermittency-uncertainty in production and behavior and difficulty to store. The other way to deal with issues of electricity generation is to manage the energy on the demand side instead of dealing with issues on the generating side. The HEMS and MEMS are efficient approaches for managing energy on the end-user side.

In traditional homes, the home is integrated with loads and receives energy from the grid. In such homes, the HEMS is only able to manage the pattern of energy consumption by shifting loads from on-peak hours when the electricity is expensive to off-peak hours when the electricity is often inexpensive.

In modern homes, the home may be integrated with renewable resources and energy storage systems and it can receive power from the external grid or send power to the grid. In such homes, the HEMS aims to change the pattern of energy consumption by shifting loads, designing an optimal charging scheduling for the energy storage system, and harvesting maximum power from renewable resources. Such homes can also operate disconnected from the grid as off-grid. In the off-grid situation, the role of the energy storage system becomes much more important because the generating energy pattern by renewables often does not match the consumed energy pattern by loads and such a mismatch of energy needs to be handled by an energy storage system.

There is a parametric uncertainty in the output power of wind and solar systems. The uncertainty can be defined as a set of scenarios of performance. Together with increasing the number of scenarios, the broader area of uncertainty is covered and the problem becomes close to the actual operation of the system. However, increasing the number of scenarios results in a complicated mathematical model with more simulation time. When the scenarios are included in the model, the power is different from one scenario to another one and it makes the problem very complicated. Such uncertainty entirely changes HEMS and must be dealt with properly. The solar, wind and load energies all can be affected by uncertainty. In such a situation, the HEMS needs to use uncertainty management systems like stochastic programming or robust programming to deal with uncertainty in the parameters.

The outage indicates the events that happen in the future because of natural disasters and attacks which lead to widespread blackouts. Renewable energies are broadly affected by events their outage is likely to happen. Such outages need to be considered and modeled by HEMS.

References

[1] Eurostat data. https://ec.europa.eu/eurostat/data/database.
[2] Hemmati R. Technical and economic analysis of home energy management system incorporating small-scale wind turbine and battery energy storage system. Journal of Cleaner Production 2017;159:106−18.

[3] Aljabery AAM, Mehrjerdi H, Mahdavi S, Hemmati R. Multi carrier energy systems and energy hubs: comprehensive review, survey and recommendations. International Journal of Hydrogen Energy 2021;46(46):23795−814.

[4] Mahapatra B, Nayyar A. Home energy management system (HEMS): concept, architecture, infrastructure, challenges and energy management schemes. Energy Systems 2022;13(3):643−69.

[5] Tostado-Véliz M, Arévalo P, Kamel S, Zawbaa HM, Jurado F. Home energy management system considering effective demand response strategies and uncertainties. Energy Reports 2022;8:5256−71.

[6] Mehrjerdi H, Bornapour M, Hemmati R, Ghiasi SMS. Unified energy management and load control in building equipped with wind-solar-battery incorporating electric and hydrogen vehicles under both connected to the grid and islanding modes. Energy. 2019;168:919−30.

[7] Mehrjerdi H, Hemmati R, Shafie-khah M, Catalão JPS. Zero energy building by multi-carrier energy systems including hydro, wind, solar, and hydrogen. IEEE Transactions on Industrial Informatics 2021;17(8):5474−84.

[8] Gao Y, Matsunami Y, Miyata S, Akashi Y. Operational optimization for off-grid renewable building energy system using deep reinforcement learning. Applied Energy 2022;325:119783.

[9] Gong H, Rallabandi V, Ionel DM, Colliver D, Duerr S, Ababei C. Dynamic modeling and optimal design for net zero energy houses including hybrid electric and thermal energy storage. IEEE Transactions on Industry Applications 2020;56(4):4102−13.

[10] Li T, Liu X, Lin Z, Morrison R. Ensemble offshore wind turbine power curve modelling − an integration of isolation forest, fast radial basis function neural network, and metaheuristic algorithm. Energy. 2022;239:122340.

[11] Faraji H, Hemmati R. A novel resilient concept for district energy system based on central battery and decentral hybrid generating resources. International Journal of Energy Research 2022;46(9):11925−42.

[12] Bornapour M, Hooshmand R-A, Khodabakhshian A, Parastegari M. Optimal stochastic coordinated scheduling of proton exchange membrane fuel cell-combined heat and power, wind and photovoltaic units in micro grids considering hydrogen storage. Applied Energy 2017;202(Suppl. C):308−22.

[13] Kazemzadeh N, Ryan SM, Hamzeei M. Robust optimization vs. stochastic programming incorporating risk measures for unit commitment with uncertain variable renewable generation. Energy Systems 2019;10(3):517−41.

[14] Abdelmalak M, Benidris M. Proactive generation redispatch to enhance power system resilience during hurricanes considering unavailability of renewable energy sources. IEEE Transactions on Industry Applications 2022;58(3):3044−53.

[15] Hemmati R, Faraji H. Identification of cyber-attack/outage/fault in zero-energy building with load and energy management strategies. Journal of Energy Storage 2022;50:104290.

[16] Mehrjerdi H, Hemmati R, Mahdavi S, Shafie-khah M, Catalao JP. Multi-carrier microgrid operation model using stochastic mixed integer linear programming. IEEE Transactions on Industrial Informatics 2022;18(7):4674−87.

[17] Hemmati R, Mehrjerdi H, Nosratabadi SM. Resilience-oriented adaptable microgrid formation in integrated electricity-gas system with deployment of multiple energy hubs. Sustainable Cities and Society 2021;71:102946.

[18] Zhou B, Xu D, Li C, Chung CY, Cao Y, Chan KW, et al. Optimal scheduling of biogas−solar−wind renewable portfolio for multicarrier energy supplies. IEEE Transactions on Power Systems 2018;33(6):6229−39.

[19] Olsen DJ, Zhang N, Kang C, Ortega-Vazquez MA, Kirschen DS. Planning low-carbon campus energy hubs. IEEE Transactions on Power Systems 2018;34(3):1895−907.

[20] Wang X, Bie Z, Liu F, Kou Y, Jiang L. Bi-level planning for integrated electricity and natural gas systems with wind power and natural gas storage. International Journal of Electrical Power & Energy Systems 2020;118:105738.
[21] Wang Y, Wang Y, Huang Y, Yang J, Ma Y, Yu H, et al. Operation optimization of regional integrated energy system based on the modeling of electricity-thermal-natural gas network. Applied Energy 2019;251:113410.
[22] Liu T, Zhang D, Wang S, Wu T. Standardized modelling and economic optimization of multi-carrier energy systems considering energy storage and demand response. Energy Conversion and Management 2019;182:126−42.
[23] Shabani MJ, Moghaddas-Tafreshi SM. Fully-decentralized coordination for simultaneous hydrogen, power, and heat interaction in a multi-carrier-energy system considering private ownership. Electric Power Systems Research 2020;180:106099.
[24] Eladl AA, El-Afifi MI, Saeed MA, El-Saadawi MM. Optimal operation of energy hubs integrated with renewable energy sources and storage devices considering CO_2 emissions. International Journal of Electrical Power & Energy Systems 2020;117:105719.

Integration of renewable resources and energy storage systems

© 2024 Elsevier Inc. All rights reserved.

Nomenclature

Parameters and variables	Definition
s	Index of scenarios
S	Set of scenarios
t	Index of time periods
T	Set of time periods
D_{ec}	Daily energy cost ($/day)
E_b^t	Energy of battery (kWh)
E_b^0	Initial energy of battery (kWh)
E_p^t	Electricity price ($/kWh)
E_{rb}	Rated capacity of battery (kWh)
F_p^t	Fuel price for diesel generator ($/kWh)
I_{du}^t	Duration of time period (hour)
K_{load}^t	Factor for curtailed load (0 to 1)
L_{shed}^t	Penalty factor for curtailed load ($/kWh)
O_{pr}^s	Probability of occurrence for scenarios
P_{chb}^t	Charged power to battery (kW)
P_{diesel}^t	Power of diesel generator (kW)
P_{dib}^t	Discharged power from battery (kW)
P_{grid}^t	Exchanged power with grid (kW)
$P_{grid}^{s,t}$	Exchanged power with grid under uncertainty (kW)
P_{load}^t	Load power (kW)
$P_{load}^{s,t}$	Load power under uncertainty (kW)
P_{rgrid}	Capacity of line between home and grid (kW)
$P_{rdiesel}$	Rated power of diesel generator (kW)
P_{rb}	Rated power of battery (kW)
P_{solar}^t	Power of solar system (kW)
$P_{solar}^{s,t}$	Power of solar system under uncertainty (kW)
P_{wind}^t	Power of wind system (kW)
$P_{wind}^{s,t}$	Power of wind system under uncertainty (kW)
η_b	Efficiency of battery energy storage system (%)

2.1 Introduction

The home energy management system (HEMS) is an optimization program in which the loads and resources of the home are managed and optimized in order to minimize costs or maximize benefits [1]. In a building equipped with local energy resources and loads, the energy management system can be carried out on the load side, the resources side, or both of them at the same time. On the load side, the HEMS can regulate the energy and power of the loads when either there is a shortage in supplying energy or the electricity is expensive. In both cases, the loads need to be categorized into different groups and HEMS deals with each group in a different way. For instance, the energy consumption pattern of some loads cannot be changed and they operate based on a specific energy pattern. As a result, such loads cannot be managed or regulated. Some other loads can only be interrupted and shifted from one hour to another hour and cannot be curtailed. The other loads may be curtailed or regulated. As a result, the categorized groups of the loads are modeled by HEMS and the HEMS can afterward deal with each group of the loads based on the load requirement [2]. The load management by HEMS may be based on a contract with the upstream grid operator. In such a situation, load energy management assists the grid when necessary [3]. On the resources side, the HEMS aims to optimize the operating pattern for each resource. When the renewables like wind turbines, solar panels, or hydro systems are integrated into the home, the purpose is to extract maximum power from these renewable energies.

In the HEMS with renewable energies, there is always a mismatch between the generation pattern of renewables and the consumption pattern of loads. Such a mismatch can be addressed by the integration of energy storage systems like battery units. The battery energy storage system can shift the excess energy of renewables from the times when the load does not need large amounts of energy to the hours when the load requires more energy. Such energy arbitrage assists the HEMS to supply the loads either without receiving power from the external grid or with minimum power taken from the utility grid [4].

If the home is equipped with enough capacity for the energy storage system, the off-grid operation of the home is also possible [5]. In the off-grid, the load demand needs to be addressed by local resources and the energy storage system plays a key role to match the generation and consumption patterns [6]. The capability of the home for off-grid operation is very vital when the events happen in the upstream grid. If the individual homes are able to work off-grid, the resilience of the grid following events is increased and the unsupplied energy of loads is reduced significantly. During off-grid operation, if the resources and energy storage unit cannot handle the load demand, load shedding may be carried out to balance the generation and demand. The loads may be categorized as critical and noncritical loads and the noncritical loads are targeted by load shedding strategy. The HEMS must always supply the critical loads even when there are events or a shortage of energy. As a result, a proper operating pattern for resources together with an optimal charging scheduling for energy storage should be designed in order to reach such purposes [7].

In practice, the resources, loads, and grid may be faced short time outages that make significant impacts on the HEMS. The HEMS should consider short time outages happening on resources and the external grid in order to present the actual results. The operating patterns of resources and storage units may be re-scheduled during the outage and they afterward drive back to the normal operating conditions when events are cleared [8]. The N-1 contingency analysis is one of the well-known methods to cope with such situations [9].

The energy production by renewables and the energy consumption by loads do not behave exactly based on the predicted patterns. The load and renewable energy patterns are associated with unpredicted alterations and such variations create uncertainty in the problem. If the HEMS is modeled and operated excluding such uncertainty in the parameters, it would not be operative in practice. In other words, if the HEMS is optimized for the normal operating conditions of the system, it would not be optimal or even feasible when the operating condition changes. In real-world homes, there is a family of operating conditions rather than one nominal operating condition. The HEMS should handle all of those operating conditions and it should ensure a feasible operation under all operating conditions. In such systems, each possible operating condition may be defined as a scenario of performance with a specific probability of occurrence. The uncertainty in the loads and renewable resources is modeled by a set of scenarios of performance. Then, the HEMS is modeled and operated under a set of scenarios of performance and it should be feasible under all scenarios. Such programming is called stochastic programming. In stochastic programming, the uncertain parameter is modeled by a number of discrete probabilistic scenarios and the model is defined over a set of scenarios and it needs to achieve a feasible operation under every single scenario. In such models, the parameters like cost may be presented as the expected value of cost instead of the actual or true value. There are some other methods to deal with uncertainty such as robust optimization in which the range of parameters is continuous and the objective function and constraints are defined only over certain sets [1].

Concerning the reviewed points, the key topics about HEMS may be listed as the integration of renewable energies, application of energy storage systems, off-grid operation, resilience following outages and uncertainty in parameters. These points are modeled, simulated, and discussed in this chapter. All the models are expressed as mixed integer linear programming in GAMS software and solved by a CPLEX solver. The input data to the programming and the outputs and results are presented in detail. Each presented topic is separately modeled and discussed and the readers can properly follow the content to become acquainted with modeling and operation of each system.

2.1.1 Some research gaps in home energy management

There are some points that have not been adequately addressed by researchers or have been rarely and marginally investigated in the literature but those points are important and practical, and they need to be considered in future works. The outlines of some major points are summarized here.

Resolution of time periods: in the regular HEMSs, the time period (or time scale) for modeling the energy and power patterns is considered equal to 1-hour, and the day is modeled by 24 time periods each one 60 minutes. Such a 1-hour resolution does not reflect the actual and real behavior of resources-loads and it cannot model the fast dynamics of the system. In realistic conditions, renewable resources would change their produced power in minutes or even seconds. As well, the thermal and electrical loads would change their power in minutes. Some researchers have considered shorter time periods such as 30 minutes or 15 minutes, where the 24 hours in one day are modeled by 48 or 96 time periods, respectively. It is also possible to consider a multitime scale model for HEMS where the fast and slow dynamics are modeled individually. Considering shorter time periods such as 2 minutes or 5 minutes matches the real conditions and achieves more realistic results.

Dynamic behavior of resources and loads: In many energy management systems, the energy resources and loads are expressed by steady-state models where the transient and dynamic behaviors are excluded. Although the steady state models are accurate and acceptable, they do not take into account the dynamic behaviors of resources. The parameters like practical voltage-current capacity, temperature, permitted consecutive operating hours, stability, reactive power capability and dynamic rating are often neglected by steady-state models. For instance, in the steady state models, the battery is charged and discharged by the operator disregarding the battery voltage level or reactive power of the interfacing inverter. In the practical condition, the battery voltage increases while charging and it decreases during the discharge process and such actual conditions make direct impacts on the charged-discharged powers. Taking into consideration the dynamic and actual behaviors of resources and loads increases the accuracy of the model.

Battery degradation and depth of discharge: Although the battery increases the flexibility of HEMS and reduces the energy cost, the charging-discharging process of the battery decreases its lifetime as well. The deeper charging–discharging makes more positive impacts on the model but it increases the battery degradation because of the depth of discharge. As a result, if the HEMS aims to discuss the economic terms, the lifetime of assets and the degradation costs should be included.

Well-being and comfort of users: If the home is integrated with local resources, the users need to install many industrial resources such as solar panels, wind turbines, diesel generators, and energy storage systems in the buildings. Such devices need to be operated and fixed by experts and technicians. They also need electrical wiring and control systems. Those devices create noises and vibrations in the buildings. In order to install such devices in the buildings, sufficient space with a proper cooling system is required. All of those points reduce the well-being and comfort of users.

2.2 Home integrated with resources and grid

Fig. 2.1 shows a typical home connected to the upstream grid as well as integrated with the wind turbine, solar panels, loads, and battery energy storage system. The HEMS controls all resources, loads, and energy storages as well as regulates the exchanged power with the grid. Regarding the renewable energy system, the purpose is to extract maximum power and inject it into the home. As a result, these generating resources are not regulated or managed and all of their output power is injected into the system.

2.2.1 Nonrenewable resources

Non-renewablebble resources like diesel generators should be operated based on an optimal pattern because their operation includes operating costs (i.e., fuel costs and maintenance costs). The diesel generator is mostly operated as a backup resource and it is run when the other

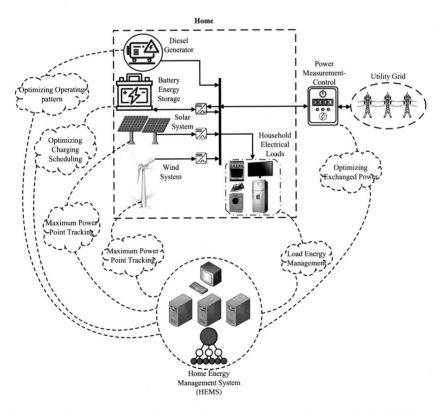

FIGURE 2.1   Home connected to the grid and integrated with wind, solar, load, and battery.

resources cannot handle the load or faced an outage. In such conditions, the HEMS has two options either curtailing the loads or utilizing the diesel generator. The decision is often made based on economic conditions. If the operating cost of the diesel generator is less than the penalty cost for load curtailment, the HEMS would run the diesel generator. Otherwise, the load-shedding strategy is applied [10].

2.2.2 Battery energy storage systems

The battery energy storage system is operated by HEMS based on optimal charging scheduling. The optimal charging—discharging operation is achieved by HEMS based on the defined objective function for optimization programming. For instance, if the economic terms are defined as the objective function, the HEMS tries to minimize the costs and utilizes battery energy storage in this way. The battery stores energy when it is low-priced and then discharges the energy when it is expensive. Such energy arbitrage properly reduces energy costs. Moreover, battery energy storage deals with the mismatch of generation and demand to satisfy the constraints of the model. If the generated power by all resources (i.e., grid, wind system, solar panels, etc.) does not match the load demand, such a mismatch makes the problem infeasible. In such situations, the energy storage system deals with the mismatch of power by charging the surplus of energy into the battery or discharging the amount of energy shortage to the home. The objective function of optimization programming in HEMS may also be included by technical terms such as unsupplied energy of loads, reliability, or resilience indexes. In such models, the HEMS attempts to minimize the curtailed or unsupplied loads by managing the resources and energy storage systems. Since such energy management is done based on technical terms, it would increase the energy cost in the home while it increases resilience and reliability on the other hand.

2.2.3 Loads and consumers

The loads are the other part of HEMS. In the residential building, there are a varied ranges of AC single-phase, AC three-phase, and DC loads. These loads can be managed, interrupted, shifted, or regulated by HEMS. The loads are categorized into various groups based on their operating manners and the HEMS copes with each group of loads in a different way. Some loads are interrupted or curtailed when either there is a shortage in supplying electricity or the energy is expensive. These loads may be connected to the electricity again when the issues are resolved [11].

2.2.4 Exchanged power with grid

The exchanged power with the external grid is also managed by HEMS. The surplus of power may be sent to the grid for making a profit. Such energy trade had better be done at hours when electricity is expensive in order to make more revenue. As a result, the energy storage system plays a key role in making a profit because it can store energy when it is cheap for selling when it is expensive. The traded power with the grid is limited by the capacity of the line installed between the network and the home. Under the limited line capacity, the HEMS needs larger local resources to supply the loads. When the capacity of the line is zero, there is an off-grid home that must supply all the loads with its own resources. In the off-grid mode, the coordinated management of resources, energy storage, and loads must be optimized to avoid load shedding or having minimum unsupplied loads.

2.3 Mathematical model for home energy management system

The HEMS with renewables, loads, battery, and the grid is mathematically formulated as a mixed integer linear programming and solved by GAMS software. A common objective function of programming is addressed by Eq. (2.1) in which the energy cost and the fuel cost are modeled and minimized. The first term shows the cost of energy that is received from the grid or sent to the external grid. The second term specifies the cost of fuel which is consumed by the diesel generator. If the HEMS does not run the diesel generator, this term is zero. This objective function may be referred to as bill cost [1].

$$D_{ec} = \sum_{t=1}^{T} \left(P^t_{grid} \times E^t_p \right) + \sum_{t=1}^{24} \left(P^t_{diesel} \times F^t_p \right) \tag{2.1}$$

In the building, the equilibrium of energy is expressed by Eq. (2.2). At every hour, the generated and consumed energies must be equal. The system is able to send power to the grid. Hence, the grid power is bidirectional and can get negative values while the other powers are positive. This point is modeled by Eq. (2.3), where the grid power is a free variable but the other variables and parameters are positive variables [1].

$$P^t_{grid} = P^t_{load} + P^t_{chb} - P^t_{dib} - P^t_{wind} - P^t_{solar} - P^t_{diesel}$$
$$\forall t \in T \tag{2.2}$$

$$\begin{cases} P^t_{load} \geq 0 \\ P^t_{chb} \geq 0 \\ P^t_{dib} \geq 0 \\ P^t_{wind} \geq 0 \\ P^t_{solar} \geq 0 \\ P^t_{diesel} \geq 0 \\ -\infty \leq P^t_{grid} \leq +\infty \end{cases} \quad \forall t \in T \tag{2.3}$$

The grid power is limited by line capacity as shown by Eq. (2.4). The diesel generator power is restrained by the rated capacity of the diesel generator as indicated in Eq. (2.5) [10]. The charged power to the battery and the discharged power from the battery is also limited by the rated power of the battery charger as modeled by Eqs. (2.6) and (2.7) [12].

$$-P_{rgrid} \leq P^t_{grid} \leq +P_{rgrid}$$
$$\forall t \in T \tag{2.4}$$

$$P^t_{diesel} \leq P_{rdiesel}$$
$$\forall t \in T \tag{2.5}$$

$$P^t_{chb} \leq P_{rb}$$
$$\forall t \in T \tag{2.6}$$

$$P^t_{dib} \leq P_{rb}$$
$$\forall t \in T \tag{2.7}$$

The battery cannot work on both the charging and discharging states at the same time. It can work either in the charging state or in the discharging mode as expressed by Eq. (2.8). The stored energy inside the battery at every hour is demonstrated by Eq. (2.9). The initial energy inside the battery at the beginning of programming can be denoted as Eq. (2.10). The efficiency of the battery energy storage system is demonstrated by Eq. (2.11) and the rated capacity of the battery is identified in Eq. (2.12) [12].

$$\begin{cases} \text{if } P^t_{chb} > 0 \Rightarrow P^t_{dib} = 0 \\ \text{if } P^t_{dib} > 0 \Rightarrow P^t_{chb} = 0 \end{cases}$$
$$\forall t \in T \tag{2.8}$$

$$E^t_b = E^{t-1}_b + \left[P^t_{chb} - \frac{P^t_{dib}}{\eta_b} \right] \times I^t_{du}$$
$$\forall t \in T, t \neq 1 \tag{2.9}$$

$$E^t_b = E^0_b + \left[P^t_{chb} - \frac{P^t_{dib}}{\eta_b} \right] \times I^t_{du}$$
$$\forall t = 1 \tag{2.10}$$

$$\eta_b = \frac{\sum_{t=1}^{T} \left(P_{dib}^t \times I_{du}^t \right)}{\sum_{t=1}^{T} \left(P_{cib}^t \times I_{du}^t \right)} \tag{2.11}$$

$$E_b^t \leq E_{rb} \tag{2.12}$$
$$\forall t \in T$$

2.4 Input data of test system

Table 2.1 shows the input data used in the proposed illustrative example. The line installed between the home and utility grid has a 20-kW capacity which is equal to peak load power. Table 2.2 presents the 24-hour patterns for solar, wind, and load powers as well as the energy price [1]. The load and renewable energy patterns in Table 2.2 are multiplied by the rated powers in Table 2.1 to achieve the pattern based on the real data [12,13].

In order to evaluate the test system, various cases are considered, simulated, and discussed. The home without resources which is a traditional home and receives its power from the grid is the first case study. The second case is the home with all the mentioned resources and it is connected to the utility grid. The off-grid operation of the home is then studied. Eventually, the uncertainties of resources and load are taken into account and the model based on stochastic programming is studied.

TABLE 2.1 The information used in the test case.

Parameter	Value
Rated power of line between grid and home (kW)	20
Rated power of diesel generator (kW)	2
Rated power of load (kW)	20
Rated power of wind system (kW)	5
Rated power of solar system (kW)	5
Rated power of battery (kW)	50
Rated capacity of battery (kWh)	50
Initial energy of battery (kWh)	0
Operating cost of diesel generator ($/kWh)	0.2

TABLE 2.2 Daily patterns of load, renewable energies and electricity price.

	Solar (p.u.)	Wind (p.u.)	Load (p.u.)	Price ($/kWh)
1	0.00	0.06	0.20	0.10
2	0.00	0.56	0.15	0.10
3	0.00	0.82	0.15	0.10
4	0.00	0.52	0.10	0.10
5	0.00	0.44	0.10	0.10
6	0.00	0.70	0.15	0.10
7	0.05	0.76	0.20	0.10
8	0.10	0.82	0.25	0.10
9	0.15	0.84	0.30	0.15
10	0.25	0.84	0.40	0.15
11	0.40	1.00	0.60	0.15
12	0.60	1.00	0.70	0.15
13	0.80	0.78	0.90	0.20
14	1.00	0.64	0.90	0.20
15	0.90	1.00	0.80	0.15
16	0.80	0.92	0.60	0.15
17	0.70	0.84	0.50	0.20
18	0.50	0.80	0.70	0.20
19	0.30	0.78	0.90	0.20
20	0.10	0.32	1.00	0.20
21	0.05	0.04	0.80	0.20
22	0.00	0.08	0.70	0.20
23	0.00	0.10	0.60	0.15
24	0.00	0.00	0.40	0.15

2.5 Operation without resources

In this case, it is assumed that the home is only connected to the grid and all the local resources (e.g., solar unit, wind turbine, and battery) are disconnected as shown in Fig. 2.2. The home has to receive all the required power of loads from the external grid. In this situation, the HEMS is modeled through Eqs. (2.13) to (2.16).

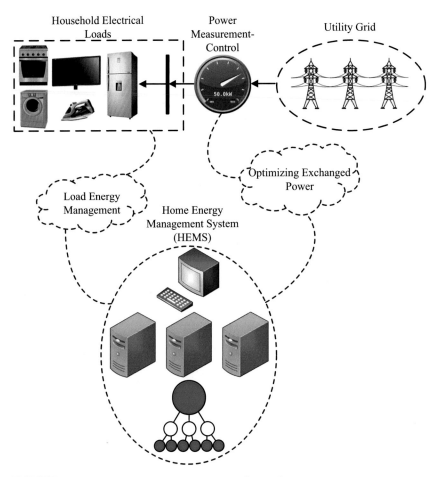

FIGURE 2.2   Home without resources connected to grid.

The objective function is modified as Eq. (2.13) and the constraints are defined through Eqs. (2.14) to (2.16).

$$D_{ec} = \sum_{t=1}^{T} \left(P_{grid}^t \times E_p^t \right) \tag{2.13}$$

$$\begin{aligned} P_{grid}^t &= P_{load}^t \\ \forall t &\in T \end{aligned} \tag{2.14}$$

$$\begin{cases} P_{load}^t \geq 0 \\ -\infty \leq P_{grid}^t \leq +\infty \end{cases} \forall t \in T \tag{2.15}$$

$$\begin{aligned} -P_{rgrid} &\leq P_{grid}^t \leq +P_{rgrid} \\ \forall t &\in T \end{aligned} \tag{2.16}$$

The simulation results of this case are listed in Table 2.3. It is clear that the received power from the grid is equal to the load power at every hour. The daily energy cost can also be calculated by using the energy prices already defined. The daily energy cost is achieved equal to 41.1 $/day. In this case, there is not any energy management in the home.

TABLE 2.3 Grid power and load power when all resources are switched off.

Hour	Grid power (kW)	Load power (kW)
1	4	4
2	3	3
3	3	3
4	2	2
5	2	2
6	3	3
7	4	4
8	5	5
9	6	6
10	8	8
11	12	12
12	14	14
13	18	18
14	18	18
15	16	16
16	12	12
17	10	10
18	14	14
19	18	18
20	20	20
21	16	16
22	14	14
23	12	12
24	8	8

2.6 Operation with all resources

In this case, the home is integrated with all the introduced resources and it is also connected to the grid. The simulation results of this case are listed in Table 2.4. It is clear that optimal energy management is carried out by HEMS. The received power from the grid is changed from

TABLE 2.4   Powers when all resources are connected.

Hour	Grid (kW)	Load (kW)	Charge power of battery (kW)	Discharge power of battery (kW)	Wind (kW)	Solar (kW)	Diesel generator (kW)
1	20	4	16.3	0	0.3	0	0
2	13.3	3	13.1	0	2.8	0	0
3	−1.1	3	0	0	4.1	0	0
4	20	2	20.6	0	2.6	0	0
5	−0.2	2	0	0	2.2	0	0
6	−0.5	3	0	0	3.5	0	0
7	−0.05	4	0	0	3.8	0.25	0
8	0.4	5	0	0	4.1	0.5	0
9	1.05	6	0	0	4.2	0.75	0
10	2.55	8	0	0	4.2	1.25	0
11	5	12	0	0	5	2	0
12	6	14	0	0	5	3	0
13	−10.1	18	0	20.2	3.9	4	0
14	−20	18	0	29.8	3.2	5	0
15	20	16	15.5	0	5	4.5	2
16	20	12	18.6	0	4.6	4	2
17	2.3	10	0	0	4.2	3.5	0
18	6	14	0	1.5	4	2.5	0
19	−20	18	0	32.6	3.9	1.5	0
20	17.9	20	0	0	1.6	0.5	0
21	15.55	16	0	0	0.2	0.25	0
22	13.6	14	0	0	0.4	0	0
23	11.5	12	0	0	0.5	0	0
24	8	8	0	0	0	0	0

hour to hour. As well, at some hours the received power from the grid is negative which shows injecting power from home to the grid. Such an operation reduces the daily energy cost significantly. The HEMS tries to send power from home to the grid when the electricity is expensive like hour 19 as well as when there is a surplus of renewable energy in the home such as hours 13−14. Sending energy to the grid makes a profit for the home resulting in bill cutting. The charged and discharged energy by the battery helps HEMS to achieve such a purpose. The battery properly shifts the produced energy of renewables to match the desired pattern. The battery charges energy at most of the hours but it discharges energy at hours 13−14 and 19−20. At these hours, the HEMS needs to send power to the grid and the battery properly assists the HEMS by releasing the stored energy. The wind and solar energies are completely consumed by the home and maximum power is extracted from them. As well, the operation of the diesel generator is not required in this case and it is seen that the diesel generator only operates at hours 15−16. The daily energy cost is 18.15 $/day which shows a great reduction compared to the previous case.

The scheduled charging operation of the battery is depicted in Fig. 2.3. Most charging operations are seen when energy is very cheap

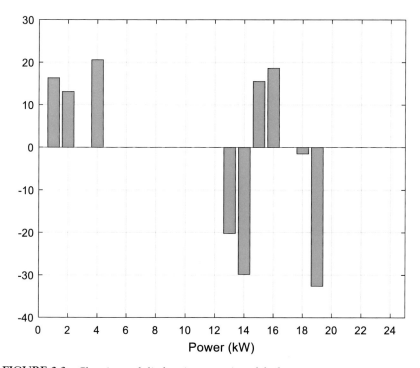

FIGURE 2.3   Charging and discharging operation of the battery.

such as the initial hours of the day or when there is a surplus of energy such as hours 15 and 16. On the other hand, the discharges happen at hours 13–14 and 19 when electricity is expensive and the HEMS is interested to send power to the grid rather than taking power from the grid. The energy stored by the battery is shown in Fig. 2.4. It is seen that the battery keeps energy until hour 12 and discharges it at hour 13. The energy is also shifted from hours 15 to 19.

As it was stated, the battery energy storage system plays a major role in the HEMS because it shifts energy over the day hours. The HEMS without battery is simulated again to evaluate the effects of battery on the model. In this case, the battery is disconnected from the home and the program is run again. The results are presented in Table 2.5. It is seen that the HEMS is not able to send energy to the external grid during hours 19–20 when the electricity is expensive. At these hours, the home receives power from the grid and has to pay the cost of expensive electricity. As well, the diesel generator has no operation in this situation. The main role of the battery is to shift energy but, in this case, there is not any device to shift energy and it makes significant effects on the energy bill. When solar and wind energies are at peak production,

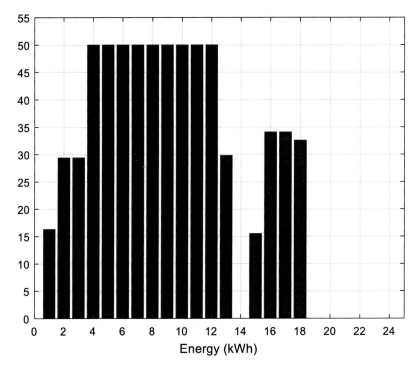

FIGURE 2.4   Energy of battery at all hours of the day.

TABLE 2.5 Powers when battery energy storage is disconnected.

Hour	Grid (kW)	Load (kW)	Wind (kW)	Solar (kW)
1	3.7	4	0.3	0
2	0.2	3	2.8	0
3	−1.1	3	4.1	0
4	−0.6	2	2.6	0
5	−0.2	2	2.2	0
6	−0.5	3	3.5	0
7	−0.05	4	3.8	0.25
8	0.4	5	4.1	0.5
9	1.05	6	4.2	0.75
10	2.55	8	4.2	1.25
11	5	12	5	2
12	6	14	5	3
13	10.1	18	3.9	4
14	9.8	18	3.2	5
15	6.5	16	5	4.5
16	3.4	12	4.6	4
17	2.3	10	4.2	3.5
18	7.5	14	4	2.5
19	12.6	18	3.9	1.5
20	17.9	20	1.6	0.5
21	15.55	16	0.2	0.25
22	13.6	14	0.4	0
23	11.5	12	0.5	0
24	8	8	0	0

the surplus of energy cannot be stored and traded in the next hours. In this case, the daily energy cost increases to 24.65 \$/day which shows 6.5 \$/day or about a 36% increase in the energy bill.

2.7 Outage of elements and resources

In the HEMS, the outage of elements and resources is likely to happen and it needs to be taken into account, evaluated, and dealt with.

Table 2.6 shows the daily energy cost following the outage of resources. The most important point is that the home continues operation when the resources are faced outage. When one element is faced with an outage, the other resources take control and compensate for its unavailability. The results demonstrate that wind energy plays a key role in supplying energy and losing wind energy results in the highest increase in the cost.

In the results shown by Table 2.6, the outage of the diesel generator does not show any impact on the energy cost but it does not mean that the diesel generator is not useful for HEMS. In order to evaluate the impacts of the diesel generator on the HEMS, the battery operation is studied when the diesel generator is faced with an outage and disconnected. Fig. 2.5 shows the charging and discharging operation of the battery when the diesel generator has faced an outage. Fig. 2.6 depicts the stored energy by the battery. Compared to the time when the diesel generator is connected, the battery shows more charging-discharging cycles. This operation reduces the battery lifetime considerably. As a result, the diesel generator does not make any impact on the energy cost directly but makes positive effects on the other resources and complements their operations.

One of the most important parts of HEMS is to schedule the power with the external grid. The capacity of the line between the home and the grid is the most important constraint on the scheduled power with the utility grid. The unlimited capacity of the line allows the HEMS to exchange boundless power with the grid but the actual current-carrying capacity is limited by the thermal rating of the line. Table 2.7 shows the energy cost under various line capacities. It is obvious that decreasing the line capacity leads to more daily energy costs because the traded power with the grid is limited and the HEMS has to utilize expensive resources like diesel generators to compensate for energy shortage. Reduction of line capacity to less than 3 kW is not possible and makes

TABLE 2.6   Daily energy cost following an outage of resources.

Disconnected elements	Daily energy cost ($/day)
NON	18.15
Battery energy storage	24.65
Diesel generator	18.15
Wind turbine	29.52
Solar system	24.42
All of 4 resources	41.10

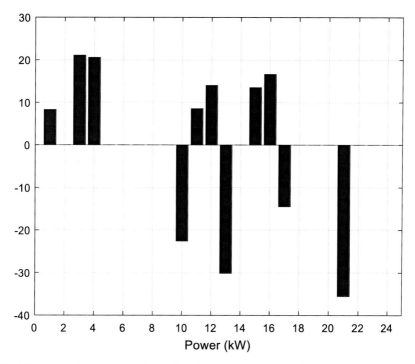

FIGURE 2.5   Charging and discharging of a battery when a diesel generator is faced outage.

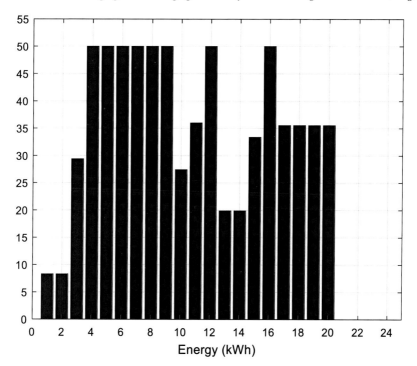

FIGURE 2.6   Energy of battery when diesel generator is faced outage.

TABLE 2.7   Energy cost under various current-carrying capacities of line.

Line capacity (kW)	Daily energy cost ($/day)
20	18.15
10	19.22
5	21.04
4	22.24
Less than or equal to 3	Infeasible operation

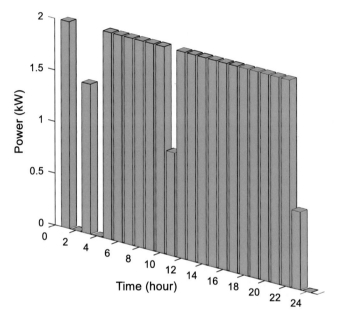

FIGURE 2.7   Produced power by diesel generator when line capacity is limited to 4 kW.

the operation infeasible. In such a situation, the resources cannot supply the load and there will be a mismatch between load and generation resulting in the infeasibility of the system.

Figs. 2.7–2.9 show the produced power by diesel generator when line capacity is limited to 4, 10, and 20 kW, respectively. The results show that the limiter line capacity results in more diesel generator operation and consequently higher energy costs for the home. In other words, when the HEMS cannot take power from the grid, it utilizes the local resources at a higher cost.

Figs. 2.10 and 2.11 indicate the charging operation of the battery when the line capacity is limited to 4 and 10 kW, respectively. The impact of limited line capacity is visible in the battery operation. The limiter line

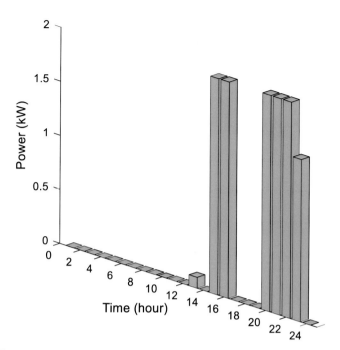

FIGURE 2.8 Produced power by diesel generator when line capacity is limited to 10 kW.

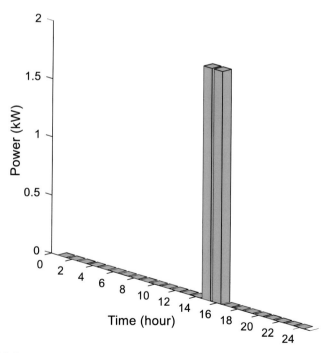

FIGURE 2.9 Produced power by diesel generator when line capacity is limited to 20 kW.

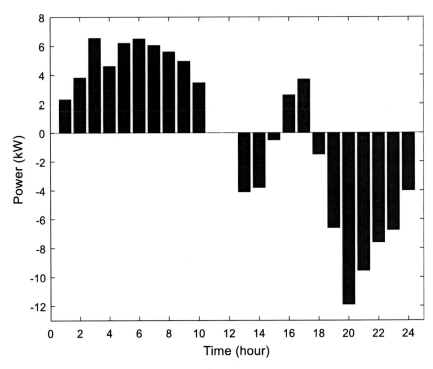

FIGURE 2.10   Operation of battery when line capacity is limited to 4 kW.

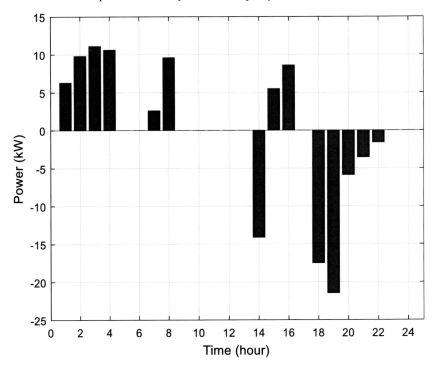

FIGURE 2.11   Operation of battery when line capacity is limited to 10 kW.

capacity forces the battery to have more operation and shift more energy. As a result, the limited line capacity increases the energy bill on the one hand and reduces the battery lifetime on the other hand.

The detailed results of HEMS, when the line capacity is limited to 4 and 10 kW, are completely presented in Tables 2.8 and 2.9. The grid and load powers, charging–discharging of battery and diesel generator

TABLE 2.8 Home energy management system when line capacity is limited to 4 kW.

Hour	Grid (kW)	Load (kW)	Charge power of battery (kW)	Discharge power of battery (kW)	Wind (kW)	Solar (kW)	Diesel generator (kW)
1	4	4	2.3	0	0.3	0	2
2	4	3	3.8	0	2.8	0	0
3	4	3	6.55	0	4.1	0	1.45
4	4	2	4.6	0	2.6	0	0
5	4	2	6.2	0	2.2	0	2
6	4	3	6.5	0	3.5	0	2
7	4	4	6.05	0	3.8	0.25	2
8	4	5	5.6	0	4.1	0.5	2
9	4	6	4.95	0	4.2	0.75	2
10	4	8	3.45	0	4.2	1.25	2
11	4	12	0	0	5	2	1
12	4	14	0	0	5	3	2
13	4	18	0	4.1	3.9	4	2
14	4	18	0	3.8	3.2	5	2
15	4	16	0	0.5	5	4.5	2
16	4	12	2.6	0	4.6	4	2
17	4	10	3.7	0	4.2	3.5	2
18	4	14	0	1.5	4	2.5	2
19	4	18	0	6.6	3.9	1.5	2
20	4	20	0	11.9	1.6	0.5	2
21	4	16	0	9.55	0.2	0.25	2
22	4	14	0	7.6	0.4	0	2
23	4	12	0	6.75	0.5	0	0.75
24	4	8	0	4	0	0	0

TABLE 2.9 Home energy management system when line capacity is limited to 10 kW.

Hour	Grid (kW)	Load (kW)	Charge power of battery (kW)	Discharge power of battery (kW)	Wind (kW)	Solar (kW)	Diesel generator (kW)
1	10	4	6.3	0	0.3	0	0
2	10	3	9.8	0	2.8	0	0
3	10	3	11.1	0	4.1	0	0
4	10	2	10.6	0	2.6	0	0
5	−0.2	2	0	0	2.2	0	0
6	−0.5	3	0	0	3.5	0	0
7	2.55	4	2.6	0	3.8	0.25	0
8	10	5	9.6	0	4.1	0.5	0
9	1.05	6	0	0	4.2	0.75	0
10	2.55	8	0	0	4.2	1.25	0
11	5	12	0	0	5	2	0
12	6	14	0	0	5	3	0
13	10	18	0	0	3.9	4	0.1
14	−4.3	18	0	14.1	3.2	5	0
15	10	16	5.5	0	5	4.5	2
16	10	12	8.6	0	4.6	4	2
17	2.3	10	0	0	4.2	3.5	0
18	−10	14	0	17.5	4	2.5	0
19	−8.85	18	0	21.45	3.9	1.5	0
20	10	20	0	5.9	1.6	0.5	2
21	10	16	0	3.55	0.2	0.25	2
22	10	14	0	1.6	0.4	0	2
23	10	12	0	0	0.5	0	1.5
24	8	8	0	0	0	0	0

operation can be evaluated at every hour of the day. Investigating the grid power reveals when the line capacity is limited to 4 kW, the home only receives power from the grid and cannot inject any power to the grid; but when the line capacity is limited to 10 kW, the home can inject power to the upstream grid at some hours like 18−19.

2.8 Off-grid operation

In this case, the home is disconnected from the grid and the exchanged power with the grid is set at zero as shown in Fig. 2.12. In order to model this situation, the grid power is re-formulated as Eq. (2.17).

$$\begin{cases} P_{rgrid} = 0 \\ -P_{rgrid} \leq P_{grid}^t \leq +P_{rgrid} \\ \forall t \in T \end{cases} \qquad (2.17)$$

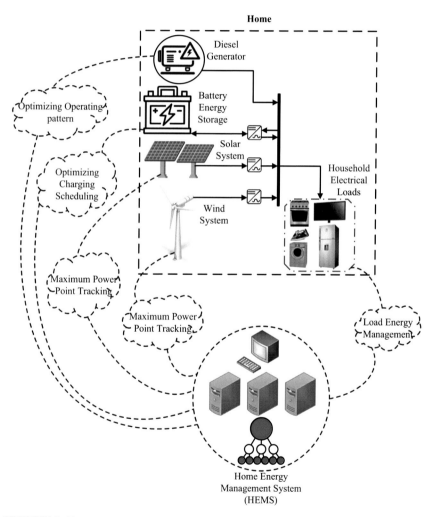

FIGURE 2.12   Off-grid model of home.

As already discussed, the home cannot continue operation when the exchanged power by the grid is less than 3 kW. As a result, in order to realize a feasible off-grid operation, the off-grid home needs larger energy resources. In this regard, the capacity of the diesel generator is increased from 2 to 6 kW. Table 2.10 demonstrates the HEMS in the

TABLE 2.10 Home energy management system in an off-grid home with 6 kW diesel generator.

Hour	Grid (kW)	Load (kW)	Charge power of battery (kW)	Discharge power of battery (kW)	Wind (kW)	Solar (kW)	Diesel generator (kW)
1	0	4	2.3	0	0.3	0	6
2	0	3	5.8	0	2.8	0	6
3	0	3	7.1	0	4.1	0	6
4	0	2	6.6	0	2.6	0	6
5	0	2	6.2	0	2.2	0	6
6	0	3	6.5	0	3.5	0	6
7	0	4	6.05	0	3.8	0.25	6
8	0	5	5.6	0	4.1	0.5	6
9	0	6	3.85	0	4.2	0.75	4.9
10	0	8	0	0	4.2	1.25	2.55
11	0	12	0	0	5	2	5
12	0	14	0	0	5	3	6
13	0	18	0	4.1	3.9	4	6
14	0	18	0	3.8	3.2	5	6
15	0	16	0	0.5	5	4.5	6
16	0	12	2.6	0	4.6	4	6
17	0	10	0.45	0	4.2	3.5	2.75
18	0	14	0	1.5	4	2.5	6
19	0	18	0	6.6	3.9	1.5	6
20	0	20	0	11.9	1.6	0.5	6
21	0	16	0	9.55	0.2	0.25	6
22	0	14	0	7.6	0.4	0	6
23	0	12	0	5.5	0.5	0	6
24	0	8	0	2	0	0	6

off-grid home with a 6-kW diesel generator. The results verify that the grid power is zero at all hours which shows off-grid operation. The diesel generator also works on the maximum power at many hours of the day. The daily energy cost is 27.04 \$/day.

As it was stated before, the home needs larger resources for off-grid operation but it is also possible to use the load-shedding strategy instead of installing larger resources. When the larger resources are not accessible, load energy management can be conducted to deal with the mismatch between generation and demand. In order to model the load-shedding strategy, the objective function and power equilibrium are re-formulated as Eqs. (2.18) and (2.19). The load shedding coefficient is also defined by Eq. (2.20). The penalty cost for load shedding is assumed 5 \$/kWh.

$$D_{ec} = \sum_{t=1}^{T} \left(P_{grid}^t \times E_p^t \right) + \sum_{t=1}^{24} \left(P_{diesel}^t \times F_p^t \right) + \sum_{t=1}^{24} \left(\left[1 - K_{load}^t \right] \times L_{shed}^t \right)$$

(2.18)

$$P_{grid}^t = \left[P_{load}^t \times K_{load}^t \right] + P_{chb}^t - P_{dib}^t - P_{wind}^t - P_{solar}^t - P_{diesel}^t$$
$$\forall t \in T$$

(2.19)

$$K_{load}^t \leq 1$$
$$\forall t \in T$$

(2.20)

Fig. 2.13 shows the load shedding in the home, where, the original load, the supplied load after load shedding, and the load shedding hours are depicted. The HEMS curtails the load when there is a mismatch between generation and load demand. Most load shedding is applied at peak-load hours when the resources cannot supply the load anymore. Table 2.11 lists the HEMS outputs under off-grid operation with load shedding strategy. The results confirm that the diesel generator operates at the maximum capacity at most hours but its operation is not enough to supply the loads and the HEMS has to carry out the load shedding and curtail some parts of the loads.

Fig. 2.14 shows the energy of the battery under the load-shedding strategy. This result indicates that the battery shifts energy from initial hours to hours 9–12 and then reallocates energy again from hours 13–16 to 17–19.

2.9 Uncertainty in parameters

The load, wind, and solar energies often come with uncertainty in behavior and their powers are not based on the predicted and predefined patterns. In order to simulate the uncertainty, a set of scenarios of performance are defined and the programming is solved under all scenarios of performance. The objective function is rearranged as Eq. (2.21),

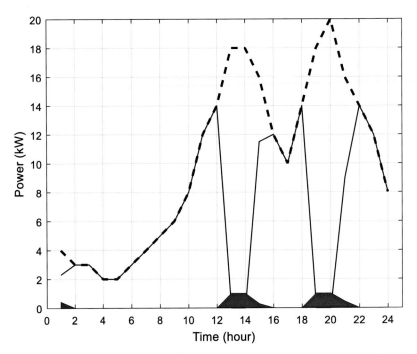

FIGURE 2.13   Load shedding in the home. Dark dashed line: original load; Blue solid line: supplied load after load shedding; Shaded area: load shedding hours.

TABLE 2.11   Home energy management system in off-grid home with load shedding.

Hour	Grid (kW)	Load (kW)	Charge power of battery (kW)	Discharge power of battery (kW)	Wind (kW)	Solar (kW)	Diesel generator (kW)
1	0	2.3	0	0	0.3	0	2
2	0	3	1.8	0	2.8	0	2
3	0	3	3.1	0	4.1	0	2
4	0	2	2.6	0	2.6	0	2
5	0	2	2.2	0	2.2	0	2
6	0	3	2.5	0	3.5	0	2
7	0	4	2.05	0	3.8	0.25	2
8	0	5	1.6	0	4.1	0.5	2
9	0	6	0.95	0	4.2	0.75	2
10	0	8	0	0.55	4.2	1.25	2
11	0	12	0	3	5	2	2

(Continued)

TABLE 2.11 (Continued)

Hour	Grid (kW)	Load (kW)	Charge power of battery (kW)	Discharge power of battery (kW)	Wind (kW)	Solar (kW)	Diesel generator (kW)
12	0	14	0	4	5	3	2
13	0	0	9.9	0	3.9	4	2
14	0	0	10.2	0	3.2	5	2
15	0	11.5	0	0	5	4.5	2
16	0	12	0	1.4	4.6	4	2
17	0	10	0	0.3	4.2	3.5	2
18	0	14	0	5.5	4	2.5	2
19	0	0	7.4	0	3.9	1.5	2
20	0	0	4.1	0	1.6	0.5	2
21	0	9	0	6.55	0.2	0.25	2
22	0	14	0	11.6	0.4	0	2
23	0	12	0	9.5	0.5	0	2
24	0	8	0	6	0	0	2

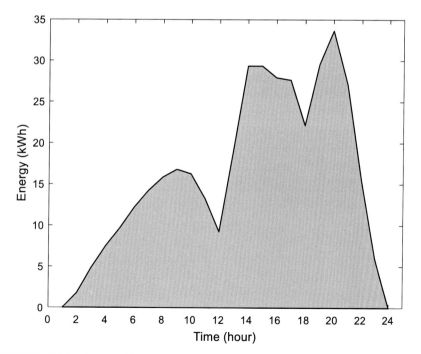

FIGURE 2.14 Energy of battery under load shedding strategy.

where, it shows the expected value of daily energy cost instead of the true value. The power balance is re-formulated as Eq. (2.22) and the parameters and variables are defined by Eq. (2.23). The grid power is rearranged as Eq. (2.24). The rest of the relationships are similar to the model without uncertainty.

$$D_{ec} = \sum_{s=1}^{S} \left[\left(\sum_{t=1}^{T} \left(P_{grid}^{s,t} \times E_p^t \right) \right) \times O_{pr}^s \right] + \sum_{t=1}^{24} \left(P_{diesel}^t \times F_p^t \right) \tag{2.21}$$

$$\begin{aligned} P_{grid}^{s,t} = P_{load}^{s,t} + P_{chb}^t - P_{dib}^t - P_{wind}^{s,t} - P_{solar}^{s,t} - P_{diesel}^t \\ \forall t \in T \\ \forall s \in S \end{aligned} \tag{2.22}$$

$$\begin{cases} P_{load}^{s,t} \geq 0 \\ P_{chb}^t \geq 0 \\ P_{dib}^t \geq 0 \\ P_{wind}^{s,t} \geq 0 \\ P_{solar}^{s,t} \geq 0 \\ P_{diesel}^t \geq 0 \\ -\infty \leq P_{grid}^{s,t} \leq +\infty \end{cases}$$
$$\begin{aligned} \forall t \in T \\ \forall s \in S \end{aligned} \tag{2.23}$$

$$\begin{aligned} -P_{rgrid} \leq P_{grid}^{s,t} \leq +P_{rgrid} \\ \forall t \in T \\ \forall s \in S \end{aligned} \tag{2.24}$$

In order to evaluate the impacts of uncertainty on the HEMS, seven scenarios of performance are defined and the model is simulated considering these scenarios. Tables 2.12–2.14 show wind, solar, and load powers

TABLE 2.12   Wind power scenarios per per-unit.

Hour	Scenario number						
	1	2	3	4	5	6	7
1	0.06	0.08	0.03	0	0.19	0.09	0
2	0.56	0.51	0.62	0.75	0.29	0.47	0.81
3	0.82	0.78	0.74	0.73	0.79	0.64	0.96
4	0.52	0.53	0.33	0.74	0.15	0.15	0.9
5	0.44	0.49	0.45	0.32	0.55	0.48	0.17
6	0.7	0.61	0.66	0.53	0.84	0.62	0.43
7	0.76	0.85	0.79	0.9	0.62	0.84	0.85

(Continued)

TABLE 2.12   (Continued)

Hour	Scenario number						
	1	2	3	4	5	6	7
8	0.82	0.78	0.65	0.62	0.88	0.46	0.2
9	0.84	0.87	0.91	0.32	0.87	1	0.1
10	0.84	0.89	0.93	0.91	1	0.96	0.84
11	1	1	0.87	0.98	0.5	0.61	0.93
12	1	0.97	0.85	0.95	0.4	1	0.1
13	0.78	0.78	0.83	0.5	0.76	0.63	1
14	0.64	0.63	0.67	0.58	0.51	0.72	0.91
15	1	0.97	0.91	0.7	0.45	1	1
16	0.92	0.93	1	0.78	0.88	0.7	0.75
17	0.84	0.86	0.84	0.72	0.8	0.12	0.44
18	0.8	0.78	0.45	0.71	0.79	0.43	0.65
19	0.78	0.55	0.54	0.25	0.61	0.69	0.08
20	0.32	0.45	0.4	0.26	0.29	0.17	0.55
21	0.04	0.04	0.19	0	0.28	0.14	0.32
22	0.08	0.12	0.05	0.01	0	0	0.26
23	0.1	0.08	0	0.21	0	0.27	0.51
24	0	0	0	0	0	0	0

TABLE 2.13   Solar power scenarios per per-unit.

Hour	Scenario number						
	1	2	3	4	5	6	7
1	0	0	0	0	0	0	0
2	0	0	0	0	0	0	0
3	0	0	0	0	0	0	0
4	0	0	0	0	0	0	0
5	0	0	0	0	0	0	0
6	0	0	0	0	0	0	0
7	0.05	0.06	0.13	0.1	0.15	0	0

(*Continued*)

TABLE 2.13 (Continued)

Hour	Scenario number						
	1	2	3	4	5	6	7
8	0.1	0.11	0.09	0.14	0.02	0	0.13
9	0.15	0.12	0.15	0.12	0.23	0.35	0.08
10	0.25	0.21	0.24	0.23	0.47	0.2	0
11	0.4	0.44	0.31	0.38	0	0.56	0
12	0.6	0.58	0.21	0.18	0.34	0.39	0.59
13	0.8	0.73	0.6	0.6	0.67	0	0.66
14	1	0.8	0.15	0.4	0.1	0.55	0.55
15	0.9	0.91	0.88	0.71	1	1	0.65
16	0.8	0.78	0.89	0.57	0.51	1	0.25
17	0.7	0.73	0.73	0.66	0.55	0.33	0.54
18	0.5	0.51	0.32	0.27	0.1	0.21	0.55
19	0.3	0.29	0.39	0.05	0.26	0.2	0
20	0.1	0.1	0	0.05	0.05	0.16	0.1
21	0.05	0.07	0.02	0	0.12	0.35	0
22	0	0	0	0	0	0	0
23	0	0	0	0	0	0	0
24	0	0	0	0	0	0	0

TABLE 2.14 Load power scenarios per per-unit.

Hour	Scenario number						
	1	2	3	4	5	6	7
1	0.2	0.22	0.22	0.06	0	0.47	0.07
2	0.15	0.11	0.25	0.11	0	0.31	0.01
3	0.15	0.16	0.14	0	0.47	0.31	0
4	0.1	0.2	0.09	0.34	0.16	0.58	0
5	0.1	0.14	0.23	0.16	0	0	0.28
6	0.15	0.2	0.3	0.25	0.37	0.33	0.12
7	0.2	0.32	0.2	0.06	0	0	0.43

(Continued)

TABLE 2.14   (Continued)

Hour	Scenario number						
	1	2	3	4	5	6	7
8	0.25	0.31	0.42	0.26	0.23	0.48	0.45
9	0.3	0.32	0.35	0.55	0.45	0.67	0.41
10	0.4	0.5	0.44	0.89	0.23	0.33	0.63
11	0.6	0.55	0.82	0.65	0.53	0.17	1
12	0.7	0.67	0.85	0.52	0.71	0.75	0.87
13	0.9	1	0.99	0.92	1	0.84	1
14	0.9	0.92	0.94	1	0.99	1.2	0.89
15	0.8	0.8	0.75	0.59	0.72	0.51	1.2
16	0.6	0.62	0.59	0.77	0.55	0.41	0.45
17	0.5	0.49	0.68	0.87	0.44	0.44	0.6
18	0.7	0.69	0.8	0.47	0.96	0.29	0.37
19	0.9	0.99	0.99	0.9	0.84	0.85	1
20	1	0.97	0.98	0.99	1.1	1.2	1.3
21	0.8	0.75	0.81	0.91	0.85	0.71	0.83
22	0.7	0.7	0.52	0.69	0.68	0.53	0.79
23	0.6	0.57	0.65	0.58	0.76	0.68	0.61
24	0.4	0.47	0.47	0.4	0.59	0.68	0.39

TABLE 2.15   Probability of scenarios of performance.

Scenario no.	Probability
1	0.20
2	0.20
3	0.15
4	0.15
5	0.10
6	0.10
7	0.10
Sum of probabilities	1

scenarios, respectively. Table 2.15 represents the probability of scenarios of performance and the sum of all probabilities is equal to one.

The proposed stochastic HEMS is simulated taking into account the defined scenarios of performance. The total energy cost is 23.011 $/day and it should be pointed out that the cost shows the expected value of the cost. Compared to the system without uncertainty, the uncertainty has

TABLE 2.16 Results of stochastic home energy management system under scenario 1.

Hour	Grid (kW)	Load (kW)	Charge power of battery (kW)	Discharge power of battery (kW)	Wind (kW)	Solar (kW)	Diesel generator (kW)
1	14.75	4	11.05	0	0.3	0	0
2	16.35	3	16.15	0	2.8	0	0
3	−1.1	3	0	0	4.1	0	0
4	−0.6	2	0	0	2.6	0	0
5	−0.2	2	0	0	2.2	0	0
6	6.65	3	7.15	0	3.5	0	0
7	15.6	4	15.65	0	3.8	0.25	0
8	0.4	5	0	0	4.1	0.5	0
9	1.05	6	0	0	4.2	0.75	0
10	2.55	8	0	0	4.2	1.25	0
11	5	12	0	0	5	2	0
12	6	14	0	0	5	3	0
13	−9.5	18	0	17.6	3.9	4	2
14	7.8	18	0	0	3.2	5	2
15	10.75	16	4.25	0	5	4.5	0
16	14.75	12	13.35	0	4.6	4	2
17	−19.55	10	0	19.85	4.2	3.5	2
18	−13.9	14	0	21.4	4	2.5	0
19	10.6	18	0	0	3.9	1.5	2
20	15.15	20	0	0.75	1.6	0.5	2
21	13.55	16	0	0	0.2	0.25	2
22	3.6	14	0	8	0.4	0	2
23	11.5	12	0	0	0.5	0	0
24	8	8	0	0	0	0	0

TABLE 2.17   Results of stochastic home energy management system under scenario 7.

Hour	Grid (kW)	Load (kW)	Charge power of battery (kW)	Discharge power of battery (kW)	Wind (kW)	Solar (kW)	Diesel generator (kW)
1	12.45	1.4	11.05	0	0	0	0
2	12.3	0.2	16.15	0	4.05	0	0
3	−4.8	0	0	0	4.8	0	0
4	−4.5	0	0	0	4.5	0	0
5	4.75	5.6	0	0	0.85	0	0
6	7.4	2.4	7.15	0	2.15	0	0
7	20	8.6	15.65	0	4.25	0	0
8	7.35	9	0	0	1	0.65	0
9	7.3	8.2	0	0	0.5	0.4	0
10	8.4	12.6	0	0	4.2	0	0
11	15.35	20	0	0	4.65	0	0
12	13.95	17.4	0	0	0.5	2.95	0
13	−7.9	20	0	17.6	5	3.3	2
14	8.5	17.8	0	0	4.55	2.75	2
15	20	24	4.25	0	5	3.25	0
16	15.35	9	13.35	0	3.75	1.25	2
17	−14.75	12	0	19.85	2.2	2.7	2
18	−20	7.4	0	21.4	3.25	2.75	0
19	17.6	20	0	0	0.4	0	2
20	20	26	0	0.75	2.75	0.5	2
21	13	16.6	0	0	1.6	0	2
22	4.5	15.8	0	8	1.3	0	2
23	9.65	12.2	0	0	2.55	0	0
24	7.8	7.8	0	0	0	0	0

increased the cost by about 27%. The results of stochastic HEMS under two scenarios 1 and 7 are presented in Tables 2.16 and 2.17. The operating patterns of the diesel generator and battery are isolated from uncertainty and they have a unique operating pattern under all uncertainties introduced by scenarios of performance. Such a point is practical because, in the actual situation, it would not be possible and practical to change the charging–discharging pattern of the battery or operating pattern of diesel following any change in the wind, solar, or load powers. As a result, a unique operation pattern is optimized for the battery and diesel generator which does not change from one scenario of performance to another one. The uncertainties in the energies and loads are handled by the upstream grid. In other words, the exchanged power with the grid changes from one scenario to another scenario.

The operating pattern of the battery under all seven scenarios is depicted in Fig. 2.15 and it is clear that the battery shows a robust pattern that is feasible under all uncertainties in renewable energies and loads. The operating pattern of the diesel generator following all changes in wind, solar, or load powers is also depicted in Fig. 2.16. The diesel generator also shows a unique operation pattern that does not change from one scenario of performance to another one.

Fig. 2.17 presents the exchanged power with the grid under all seven scenarios in stochastic HEMS. It is clear that the grid power changes under various scenarios and it deals with load and renewable energy uncertainty. The mismatch between load and generation is handled by the grid. The detailed powers of the grid are listed in Table 2.18.

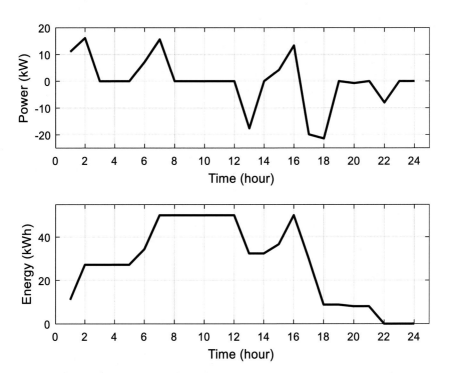

FIGURE 2.15   Charging scheduling and energy of battery under all seven scenarios.

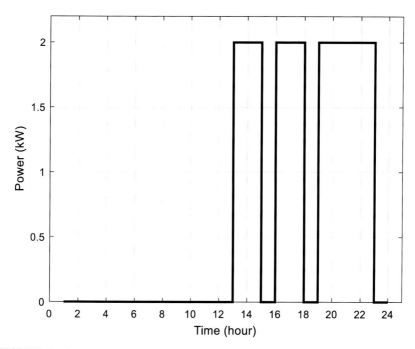

FIGURE 2.16 Operating pattern of diesel generator under all seven scenarios.

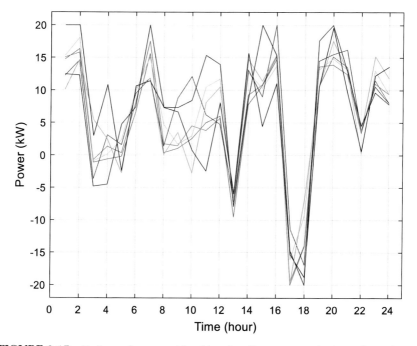

FIGURE 2.17 Exchanged power with grid under all seven scenarios in stochastic home energy management system.

TABLE 2.18 Detailed grid powers per kW under all seven scenarios.

Hour	Scenario number						
	1	2	3	4	5	6	7
1	14.75	15.05	15.3	12.25	10.1	20	12.45
2	16.35	15.8	18.05	14.6	14.7	20	12.3
3	−1.1	−0.7	−0.9	−3.65	5.45	3	−4.8
4	−0.6	1.35	0.15	3.1	2.45	10.85	−4.5
5	−0.2	0.35	2.35	1.6	−2.75	−2.4	4.75
6	6.65	8.1	9.85	9.5	10.35	10.65	7.4
7	15.6	17.5	15.05	11.85	11.8	11.45	20
8	0.4	1.75	4.7	1.4	0.1	7.3	7.35
9	1.05	1.45	1.7	8.8	3.5	6.65	7.3
10	2.55	4.5	2.95	12.1	−2.75	0.8	8.4
11	5	3.8	10.5	6.2	8.1	−2.45	15.35
12	6	5.65	11.7	4.75	10.5	8.05	13.95
13	−9.5	−7.15	−6.95	−6.7	−6.75	−5.95	−7.9
14	7.8	9.25	12.7	13.1	14.75	15.65	8.5
15	10.75	10.85	10.3	9	11.4	4.45	20
16	14.75	15.2	13.7	20	15.4	11.05	15.35
17	−19.55	−20	−16.1	−11.35	−19.8	−15.3	−14.75
18	−13.9	−14.05	−9.25	−16.9	−6.65	−18.8	−20
19	10.6	13.6	13.15	14.5	10.45	10.55	17.6
20	15.15	13.9	14.85	15.5	17.55	19.6	20
21	13.55	12.45	13.15	16.2	13	9.75	13
22	3.6	3.4	0.15	3.75	3.6	0.6	4.5
23	11.5	11	13	10.55	15.2	12.25	9.65
24	8	9.4	9.4	8	11.8	13.6	7.8

It should be noted that if the grid is not connected and the home is modeled as off-grid, one of the resources (e.g., battery) needs to handle such uncertainty. Otherwise, there will be a mismatch between load and generation that makes the HEMS infeasible.

The equilibrium of generation and consumption needs to be passed under all scenarios of performance at all hours of the day. Figs. 2.18 and

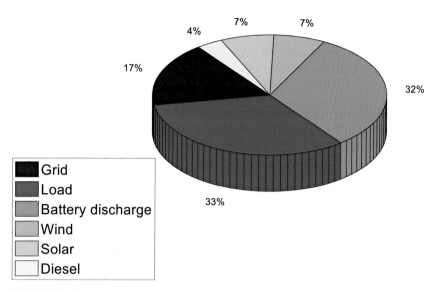

FIGURE 2.18   Generated and consumed powers at hour 13 under scenario 1.

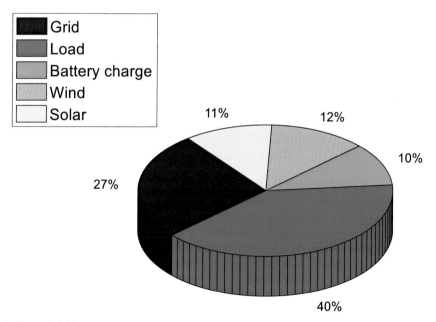

FIGURE 2.19   Generated and consumed powers at hour 15 under scenario 1.

2.19 show the generation and consumption at hours 13 and 15, respectively. At hour 13, the battery is in the discharging state and generates
electricity. The wind, solar, and diesel units also generate electricity.

The generated energy by these four resources is consumed by the loads and grid, where, 66% is consumed by the load and 34% is injected into the grid. The battery, diesel generator, wind, and solar units are generating electricity but the load and grid are consuming energy.

At hour 15, the battery is in the charging state and consumes energy and the diesel generator power is zero. The wind and solar units also generate energy. The generated energy by wind and solar is not enough to supply the loads and battery. As a result, the rest of the energy is taken from the grid. The grid, wind, and solar units are generating electricity but load and battery are consuming energy.

2.10 Conclusions

In this chapter, the comprehensive modeling and simulations of home integrated with renewables, loads, battery energy storage, and grid were discussed. First, the home without resources was studied. In this situation, the total load of the home was supplied by the grid and the daily energy cost was 41.1 $/day. Then, the home integrated with the solar system, wind unit, energy storage, and grid was modeled. The objective function was the minimization of daily energy costs. The energy cost was expressed as the purchased electricity from the grid as well as the fuel cost of the diesel generator. The HEMS was carried out to minimize the cost. The HEMS sent energy from home to the grid when the electricity was expensive or when there was a surplus of energy such as hours 13–14 and 19. The battery helped HEMS to shift energy over the hours of the day. The battery stored energy when it was either cheap or there was an excess of energy. On the other hand, the battery restored energy when energy was expensive or there was a shortage in supplying energy. The battery charged energy at most of the hours but it discharged energy at hours 13–14 and 19–20. The daily energy cost was reduced to 18.15 $/day which showed a 56% reduction compared to the first case.

The HEMS without battery was simulated to demonstrate the effects of battery energy storage on the problem. It was shown that the HEMS without battery energy storage is not able to send energy to the grid during hours 19–20 when electricity is expensive. Without batteries, the produced energy by solar and wind systems cannot be stored and shifted. Such a point limited the ability of HEMS for sending power to the grid and making revenue. As well, during the hours when the electricity was expensive like hours 19–20, the home had to receive power from the grid which increased the energy bill. The daily energy cost increased to 24.65 $/day which showed a 36% increase in the energy bill.

Under the outage of resources, the HEMS could manage energy in the home and continue operation but at higher bill costs. Losing wind energy provided the highest increase in energy costs. The diesel generator assisted the battery in having fewer operating cycles which increased the battery's lifetime.

The capacity of the line between the home and the grid was studied under various line capacity ratings. When line capacity was reduced from 20 to 4 kW, the diesel generator operation increased accordingly to compensate shortage in supplying energy from the grid side. It was found that the HEMS becomes infeasible when the line capacity is reduced to less than 3 kW. Since the line capacity cannot be less than 3 kW, off-grid operation is not possible for the home. In order to realize a feasible off-grid operation, the off-grid home needed larger energy resources. As a result, the capacity of the diesel generator was increased from 2 to 6 kW. The results showed that the off-grid home with a 6-kW diesel generator has feasible operation but the diesel generator worked on the maximum power at many hours and increased the daily energy cost to 27.04 $/day.

In the off-grid, if it is not possible to integrate the larger resources, the load-shedding strategy can be used instead. According to the load-shedding strategy, the HEMS curtailed the loads when there was a mismatch between generation and load demand especially at peak hours.

In order to model the HEMS with uncertainty, the load power, wind energy, and solar energy were modeled by a set of scenarios of performance. In this situation, daily energy cost was increased to 23.011 $/day which showed about a 27% increase compared to the model without uncertainty. The battery and diesel generator operations were modeled unrelated to scenarios of performance and they had a unique operation pattern under all scenarios of performance. The mismatch between load and generation was handled by the grid and the exchanged power with the grid changed under various scenarios.

References

[1] Hemmati R, Saboori H. Stochastic optimal battery storage sizing and scheduling in home energy management systems equipped with solar photovoltaic panels. Energy and Buildings 2017;152:290−300.
[2] Yang J, Sun Q, Yao L, Liu Y, Yang T, Chu C, et al. A novel dynamic load-priority-based scheduling strategy for home energy management system. Journal of Cleaner Production 2023;389:135978.
[3] Shareef H, Ahmed MS, Mohamed A, Hassan EA. Review on home energy management system considering demand responses, smart technologies, and intelligent controllers. IEEE Access 2018;6:24498−509.
[4] Tostado-Véliz M, Hasanien HM, Turky RA, Assolami YO, Vera D, Jurado F. Optimal home energy management including batteries and heterogenous uncertainties. Journal of Energy Storage 2023;60:106646.

[5] Keskin Arabul F, Arabul AY, Kumru CF, Boynuegri AR. Providing energy management of a fuel cell–battery–wind turbine–solar panel hybrid off grid smart home system. International Journal of Hydrogen Energy 2017;42(43):26906–13.

[6] Shaterabadi M, Jirdehi MA, Amiri N, Omidi S. Enhancement the economical and environmental aspects of plus-zero energy buildings integrated with INVELOX turbines. Renewable Energy 2020;153:1355–67.

[7] Rifat Boynuegri A, Tekgun B. Real-time energy management in an off-grid smart home: flexible demand side control with electric vehicle and green hydrogen production. International Journal of Hydrogen Energy 2023.

[8] Candan AK, Boynuegri AR, Onat N. Home energy management system for enhancing grid resiliency in post-disaster recovery period using electric vehicle. Sustainable Energy, Grids and Networks 2023;34:101015.

[9] Faraji J, Hashemi-Dezaki H, Ketabi A. Stochastic operation and scheduling of energy hub considering renewable energy sources' uncertainty and N-1 contingency. Sustainable Cities and Society 2021;65:102578.

[10] Mahdavi S, Hemmati R, Jirdehi MA. Two-level planning for coordination of energy storage systems and wind-solar-diesel units in active distribution networks. Energy. 2018;151:954–65.

[11] Faraji H, Nosratabadi SM, Hemmati R. AC unbalanced and DC load management in multi-bus residential microgrid integrated with hybrid capacity resources. Energy. 2022;252:124070.

[12] Hemmati R. Optimal design and operation of energy storage systems and generators in the network installed with wind turbines considering practical characteristics of storage units as design variable. Journal of Cleaner Production 2018;185:680–93.

[13] Mehrjerdi H, Hemmati R. Coordination of vehicle-to-home and renewable capacity resources for energy management in resilience and self-healing building. Renewable Energy 2020;146:568–79.

Integration of electric vehicles and charging stations

79
© 2024 Elsevier Inc. All rights reserved.

Nomenclature

Indexes and sets

s	Index of scenarios
S	Set of scenarios
t	Index of time periods
T	Set of time periods
v	Index of electric vehicles
V	Set of electric vehicles
b	Index of batteries for swapping
B	Set of batteries for swapping

Parameters and variables

C_{cloud}^t	Penalty cost of load curtailment (\$/kWh)
D_{ec}	Daily energy cost (\$/day)
$E_{bs}^{t,b}$	Energy of battery in swapping station (kWh)
E_{bs}^r	Rated capacity of battery in swapping station (kWh)
$E_{ev}^{t,v}$	Energy of electric vehicle (kWh)
$E_{ev}^{r,v}$	Rated capacity of electric vehicle (kWh)
E_p^t	Electricity price (\$/kWh)
E_{sload}^r	Rated energy of shiftable load (kWh)
F_p^t	Fuel cost (\$/kWh)
I_{dur}^t	Duration of time interval (minute)
K_{cloud}^t	Factor of load curtailment
N_h	Operating hours of interruptible load
O_{pr}^s	Probability of occurrence for scenarios
P_{cp}^t	Charged power to parking station (kW)
P_{cloud}^t	Power of curtailable load (kW)
$P_{cev}^{t,v}$	Charged power to electric vehicle (kW)
P_{cbss}^t	Charged power to battery swapping station (kW)
$P_{cbs}^{t,b}$	Charged power to each battery in the swapping station (kW)
P_{diesel}^t	Power of diesel generator (kW)
Pr_{diesel}^r	Rated power of diesel generator (kW)
P_{dp}^t	Discharged power from parking station (kW)
$P_{dev}^{t,v}$	Discharged power from electric vehicle (kW)
P_{dbss}^t	Discharged power from battery swapping station (kW)
$P_{dbs}^{t,b}$	Discharged power from each battery in swapping station (kW)
$P_{ev}^{r,v}$	Rated power of electric vehicle (kW)
P_{fload}^t	Power of fixed load (kW)
P_{grid}^t	Exchanged power with grid (kW)
P_{grid}^r	Rated power of line (kW)
$P_{grid}^{t,s}$	Exchanged power with grid under uncertainty (kW)
P_{iload}^t	Power of interruptible load (kW)
P_{iload}^r	Rated power of interruptible load (kW)
P_{load}^t	Total power of load (kW)
$P_{load}^{t,s}$	Total power of load under uncertainty (kW)
P_{sload}^t	Power of shiftable load (kW)
P_{sload}^r	Rated power of shiftable load (kW)

P^t_{solar}	Power of solar system (kW)
$P^{t,s}_{solar}$	Power of solar system under uncertainty (kW)
t_{out}	Time interval when electric vehicle exiting the station
t_{swap}	Time interval when battery is swapped
$u^{t,v}_{cev}$	Binary variable showing electric vehicle charging operation
$u^{t,b}_{cbs}$	Binary variable showing charging operation of battery in swapping station
$u^{t,v}_{dev}$	Binary variable showing electric vehicle discharging operation
$u^{t,b}_{dbs}$	Binary variable showing discharging operation of battery in swapping station
u^t_{iload}	Binary variable showing interruptible load operation
η_{ev}	Efficiency of electric vehicle (%)
η_{bs}	Efficiency of battery in swapping station (%)

3.1 Introduction

Electric vehicles have been broadly developed in the recent decade and the increasing number of electric vehicles in societies has created new challenges, opportunities, and issues. Electric vehicles reduce the environmental issues which are caused by petroleum-based transportation infrastructure. The energy required for electric vehicles can be supplied by a wide range of energy resources such as fossil fuels, nuclear power, hydropower, renewable systems, or any combination of those resources but fossil fuel-powered vehicles only consume fossil fuels. The environmental pollution released by electric vehicles varies depending on the fuel and technology used to supply their energy. If they use renewable energies like solar and wind generating systems, their carbon footprint will be zero. Electric vehicles use on-board batteries or capacitors to store energy. One of the major benefits of electric vehicles is regenerative braking, which recovers kinetic energy as electricity and charges it into the on-board energy storage system [1]. The electric vehicles may utilize on-board generating systems such as fuel-cell (fuel cell vehicles) or diesel engines. The on-board energy storage system can be either charged by connection to the land-based charging stations or by an on-board power source such as an internal combustion engine. If the electric vehicle can be charged by any external source of electricity, it is called a plug-in electric vehicle. The lithium-ion batteries are the common type of energy storage system in electric vehicles because of their advantages like longer lifetime and more energy-power density. Together with developing new technologies, the prices of lithium-ion batteries are continuously decreasing, which reduces the final price of electric vehicles [2].

One of the substantial challenges facing electric vehicles is to design and install efficient and proper charging stations [3]. A charging station or charging point charges the battery of plug-in electric vehicles. The charging stations are the AC and DC types and the DC models provide higher charging power. Levels 1 to 3 are defined for charging speed. Level 1 presents a 120-V charger using the standard electrical outlet. The charging time is about 16 hours for a 130 km battery. The range per hour is about 5−8 km per hour of charge. Level 2 presents a 240-V charger with about 32−96 km per hour of charge. The charging time is about 3.5 hours for a 130 km battery. Level 3 uses 480-Vt DC with about 96−160 km per 20 minutes of charge. The charging time is about 30 minutes [4]. The charging stations are investigated from various points of view such as optimal location [5], optimal charging scheduling [6], the impacts on the utility grid [6], off-grid charging stations [7], mobile charging stations, and wireless charging systems [8].

While electric cars are attaining popularity fast, some potential customers remain hesitant because of the issues related to electric vehicles. The long charging time is one of those matters.

While the gas tank of petroleum-based cars is filled in less than five minutes, the fastest available charger of electric vehicles is able to charge 80% of the battery in about 30 minutes. The slow charging time of electric vehicles is holding back the potential electric vehicle customers and one of the key issues related to electric vehicle adoption is the duration of charging time which is more than pumping gases. For the wide adoption of electric vehicles, the charging time of electric cars needs to be as fast as pumping gases. Battery swapping is a new approach to overcome the slow charging of electric cars. The battery swapping stations present a new model for charging electric vehicles. In these stations, the depleted battery of an electric vehicle is mechanically exchanged with a fully charged battery in a few minutes. The battery swapping stations are investigated from various points of view such as sizing, siting, optimal charging scheduling, and cost-benefit analysis [9].

In this chapter, the energy management in homes and residential microgrids is addressed taking into account different models of electric vehicles. The electric vehicle parking station, electric vehicle charging station, and battery swapping station are modeled and integrated to home energy management and residential microgrid energy management. In the home energy management with the electric vehicle parking station, the test system is a home integrated with solar panels, diesel generator, loads, external grid, and electric vehicle parking station. The home energy management system (HEMS) optimizes the charging scheduling of electric vehicles, finds the operating pattern of diesel generators, harvests the maximum power of the solar system, and adjusts the operation of loads. The loads are modeled as fixed, interruptible, shiftable, and curtailable loads and their optimal management is done by HEMS. The model is investigated by taking into account the outages, parametric uncertainty and discharge capability.

In the residential microgrid energy management with electric vehicle charging station, the operation of electric vehicles is modeled based on a charging station model; where each arrived electric vehicle is charged and leaves the station opposite the parking station, where the electric vehicles are parked for several hours. The residential microgrid is integrated with solar energy, a diesel generator, an external grid, a charging station, and loads (i.e., fixed, interruptible, shiftable, and curtailable loads). The outages, events, various discharge selections, and parametric uncertainty are discussed.

The energy management in residential microgrid with battery swapping station is also modeled. The microgrid structure and resources are similar to the previous case but the microgrid is integrated with the battery swapping station rather than the charging station.

3.1.1 Some research gaps in electric vehicle energy management

There are some points that have not been adequately addressed by researchers or have been rarely and marginally investigated in the literature but those points are important and practical, and they need to be considered in future works. The outlines of some major points are summarized here.

Resolution of time-periods: in modern charging stations, the high voltage-power chargers are able to charge electric vehicles in less than 20 minutes. As a result, it is required to consider very short time-periods for energy management of charging stations. Otherwise, the electric vehicles may not be efficiently charged or discharged. In order to achieve efficient charging scheduling, shorter time-periods such as 2-minute or 5-minute should be considered.

Dynamic behavior of electric vehicle battery: The fast dynamics such as battery voltage drop during discharging process create prominent barriers and issues in the energy management of charging stations. For instance, considering the time-periods less than 1 or 2 minutes for energy management of electric vehicles may be affected by voltage drop-rise during discharging–charging operation.

Policies and regulations: Discharging the battery of an electric vehicle and using it as an energy resource leads to the degradation of the battery and reduces its lifetime. Such points must be accepted and approved by electric vehicles owners. Such a paradigm needs making rules and policies by companies or the government and both the car and station owners should comply with the rules.

Swapping new battery with old one: In the battery swapping station, the battery of a new electric vehicle that is a new battery may be replaced with an old battery which has been used for years. Such a point is not accepted by the drivers and would make a negative impact on the adoption of battery-swapping stations. The ownership of batteries in such a paradigm is still a matter.

Well-being and comfort of drivers: Keeping the electric vehicles inside the charging stations for hours is not possible and it would reduce the well-being and comfort of drivers. It is required to offer significant incentives for persuading the drivers to participate in the energy management system.

Uncertainty in the behavior of electric vehicles: The electric vehicles may arrive at the charging station with an empty battery or a battery that comprises some amount of energy. They would leave the charging station before getting fully charged. The pattern of electric vehicles that enter the charging station is not exactly predictable. The electric vehicles in the parking station would need to leave the parking before the predetermined time. Such uncertainties make significant effects on the energy management system.

3.2 HEMS with electric vehicle parking station

Fig. 3.1 shows a typical home that is integrated with the local resources, loads, external grid as well as the electric vehicle parking station. The HEMS controls and regulates all resources and electric vehicles in order to minimize the cost of traded power with the grid. The home has a parking station, where electric vehicles can be parked and charged [10]. In the parking station, each electric vehicle may be parked for several hours. During this time-period, the HEMS optimizes the charged power to the electric vehicles while the HEMS is allowed to utilize the electric vehicles as a resource by discharging power from their on-board batteries. The optimized charging-discharging process of each electric vehicle is continued until the time when the electric vehicle aims to leave the parking station. At this time, the electric vehicle must be fully charged and ready for use by the driver. As a result, an optimal charging-discharging process is designed for each electric vehicle in the parking station.

The other local resources are also optimally operated by HEMS. The solar energy system is integrated into the home and HEMS aims to harvest maximum power from this resource. As a result, all produced power by the solar system is injected into the home and the produced power is not regulated. The diesel generator is often equipped for backup operation and it is not used as a 24-hour generating system. The operating pattern of the diesel generator is needed to be optimized by the HEMS. The loads

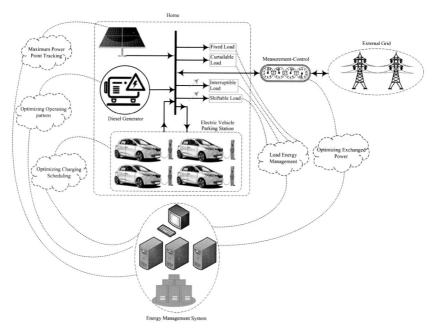

FIGURE 3.1   HEMS including resources, loads, external grid, and electric vehicle parking station.

in the home show various behaviors and characteristics. They are therefore modeled based on their actual behavior. The loads may be modeled as shift-able, interruptible, curtailable, and fixed (i.e., uninterruptible) loads. The fixed loads consume a fixed pattern of energy over 24-hour and such demand for energy cannot be adjusted, changed, or curtailed.

The home with all resources is connected to the utility grid and the HEMS aims to minimize the energy cost by optimizing operating patterns of loads, resources, and electric vehicles. The home can manage the exchanged power with the grid in two ways;

- The received power from the grid can be shifted from the hours when electricity is expensive to the hours with inexpensive energy in order to decrease the energy cost. Such operation is achieved by shifting the consumers' operation to low-priced energy hours. As well, the energy can be stored in the electric vehicles at low-priced hours and then restored when the electricity is expensive.
- The other way to reduce the bill cost is to send energy to the grid for making revenue. The energy exchange with the grid for making revenue is economical when the electricity is expensive. At high-priced energy hours, the excess energy in the home may be traded with the external grid.

During events and off-grid operation, the diesel generator can be operated to supply the load demand and avoid load curtailment. In such situations, there are two options for HEMS including load curtailment (load demand reduction) or using the diesel generator for avoiding load curtailment. The HEMS utilizes one of those options with regard to the economic and technical barriers.

3.2.1 Formulation of HEMS with parking station

The objective function of HEMS is to minimize the daily energy cost which is expressed by Eq. (3.1). The first term shows the electricity cost, the second term indicates the diesel generator operational cost and the last term shows the penalty cost for load curtailment. The HEMS uses the load curtailment option at a higher cost when the energy resources cannot supply the loads [10].

$$D_{ec} = \sum_{t=1}^{T} \left(P^t_{grid} \times E^t_p \right) + \sum_{t=1}^{T} \left(P^t_{diesel} \times F^t_p \right) + \sum_{t=1}^{T} \left(1 - K^t_{cload} \right) \times C^t_{cload} \quad (3.1)$$

The balance of power in the system is specified by Eq. (3.2). The surplus of power in the home is sent to the grid and the shortage of power is supplied by the grid. As a result, the grid power can take both positive and negative values while the other powers are positive. The capacity of the line between the home and the grid is modeled by Eq. (3.3) and the rated power of the diesel generator is modeled by Eq. (3.4).

$$P^t_{grid} = P^t_{load} + P^t_{cp} - P^t_{dp} - P^t_{solar} - P^t_{diesel} \\ \forall t \in T \quad (3.2)$$

$$\left| P^t_{grid} \right| \leq P^r_{grid} \\ \forall t \in T \quad (3.3)$$

$$P^t_{diesel} \leq P^r_{diesel} \\ \forall t \in T \quad (3.4)$$

The total load of the home is expressed by Eq. (3.5) and it has four terms including fixed loads, interruptible loads, shiftable loads, and curtailable loads. Load curtailment is an option to deal with the shortage of energy in the building. The factor of load curtailment is expressed by Eq. (3.6). The energy of shiftable load is denoted by Eq. (3.7) and its rated power is modeled by Eq. (3.8). The shiftable loads need specific energy during 24-hour and such energy can be supplied at various hours. The interruptible load is modeled by Eq. (3.9) and the consecutive hours that are required for the operation of the interruptible load are specified by Eqs. (3.10) and (3.11). The interruptible load can be interrupted and operated again in the next hours. However, such loads need to be operated for several consecutive hours to finish their task. As a result, they need to remain connected to the electricity for several hours after getting switched on [10].

$$P_{load}^t = P_{fload}^t + P_{iload}^t + P_{sload}^t + \left(P_{cload}^t \times K_{cload}^t\right)$$
$$\forall t \in T \tag{3.5}$$

$$K_{cload}^t \leq 1$$
$$\forall t \in T \tag{3.6}$$

$$\sum_{t=1}^{T} P_{sload}^t = E_{sload}^r \tag{3.7}$$

$$P_{sload}^t \leq P_{sload}^r$$
$$\forall t \in T \tag{3.8}$$

$$P_{iload}^t = P_{iload}^r \times u_{iload}^t$$
$$\forall t \in T \tag{3.9}$$

$$\sum_{t=1}^{T} u_{iload}^t = N_h$$
$$\forall t \in T \tag{3.10}$$

$$\begin{cases} u_{iload}^t = 1 \ \forall t \in [t, \cdots t + N_h] \\ u_{iload}^t = 0 \ \forall t \notin [t, \cdots t + N_h] \end{cases} \tag{3.11}$$

The electric vehicles in the parking station can be charged at various hours and they are also modeled to be discharged when necessary. The charged power to the electric vehicle and the discharged power from the electric vehicle are modeled by Eqs. (3.12) to (3.14). The energy of the electric vehicle is calculated in Eq. (3.15) and its efficiency is presented in Eq. (3.16). The rated capacity of the electric vehicle is formulated by Eq. (3.17) [11].

$$u_{cev}^{t,v} + u_{dev}^{t,v} \leq 1$$
$$\forall t \in T$$
$$\forall v = V \tag{3.12}$$

$$P_{cev}^{t,v} \leq P_{ev}^{r,v} \times u_{cev}^{t,v}$$
$$\forall t \in T$$
$$\forall v = V \tag{3.13}$$

$$P_{dev}^{t,v} \leq P_{ev}^{r,v} \times u_{dp}^{t,v}$$
$$\forall t \in T$$
$$\forall v = V \tag{3.14}$$

$$E_{ev}^{t,v} = E_{ev}^{t-1,v} + \left\{ P_{cev}^{t,v} - \frac{P_{dev}^{t,v}}{\eta_{ev}} \right\} \times I_{dur}^{t}$$

$$\forall t \in T$$
$$\forall v = V$$

(3.15)

$$\eta_{ev} = \frac{\sum_{t=1}^{T} \left(P_{dev}^{t,v} \times I_{dur}^{t} \right)}{\sum_{t=1}^{T} \left(P_{cev}^{t,v} \times I_{dur}^{t} \right)}$$

(3.16)

$$E_{ev}^{t,v} \leq E_{ev}^{r,v}$$
$$\forall t \in T$$
$$\forall v \in V$$

(3.17)

When the electric vehicle is leaving the parking station it must be fully charged and this point is indicated by Eq. (3.18). The total power charged to the parking station and the total power discharged from the parking station are calculated by Eqs. (3.19) and (3.20).

$$E_{ev}^{t,v} = E_{ev}^{r,v}$$
$$\forall t = t_{out}$$
$$\forall v = V$$

(3.18)

$$\sum_{v=1}^{V} \left(P_{cev}^{t,v} \right) = P_{cp}^{t}$$
$$\forall t \in T$$

(3.19)

$$\sum_{v=1}^{V} \left(P_{dev}^{t,v} \right) = P_{dp}^{t}$$
$$\forall t \in T$$

(3.20)

3.2.2 Case study for HEMS with parking station

The technical and economic data of the proposed test case are presented in Table 3.1. The data of lines, loads, resources, and electric vehicles are presented. Table 3.2 presents 24-hour patterns of loads, energy price, and solar energy. The presented powers are per-unit based on the rated power [7,10,11].

Table 3.3 shows the pattern of availability of electric vehicles in the parking station. The hours denoted by 1 show the availability of electric vehicles and the hours denoted by 0 show the time periods when the

TABLE 3.1 Technical and economic data of test case.

Parameter	Level
Grid line capacity	50 kW
Rated power of diesel generator	5 kW
Rated power of interruptible load	15 kW
Consecutive hours for interruptible load operation	6 hours
Rated power of curtailable load	10 kW
Rated power of fixed load	20 kW
Rated energy of shift-able load	40 kWh
Rated power of shift-able load	10 kW
Rated power of solar system	15 kW
Fuel cost of diesel generator	0.2 $/kWh
Penalty of load curtailment	5 $/kWh
Rated capacity of electric vehicle battery	50 kWh
Rated power of electric vehicle charger	50 kW
Initial energy of electric vehicle	20 kWh

electric vehicles are not in the parking station. At the last hour of availability of electric vehicles in the parking station, they must get fully charged.

3.2.3 Simulation of HEMS with parking station

The proposed optimization programming is simulated in GAMS software and the results are discussed here. The daily energy cost is achieved at 55.965 $/day. Loads of the home are optimized as listed in Table 3.4. The fixed loads are supplied at all hours without any interruption or curtailment. The interpretable load is operated for 6 consecutive hours, especially at the initial hours of the day when the electricity is inexpensive. The energy of the shiftable load is 40 kWh and it is supplied at 4 various hours. The load curtailment option is not utilized by HEMS and it is seen that the curtailed load is zero at all hours. Since there is no shortage of energy in the system, the HEMS uses electricity from the grid instead of load curtailment resulting in less energy cost for the home.

The powers of loads, resources, and grid are listed in Table 3.5. At some hours the grid power is negative which shows sending power

TABLE 3.2   Patterns of load power, energy price and solar energy.

Hour	Fixed load power (p.u.)	Curtailable load power (p.u.)	Energy price ($/kWh)	Solar power (p.u.)
1	0.2	0.1	0.1	0
2	0.15	0.1	0.1	0
3	0.15	0.1	0.1	0
4	0.1	0.08	0.1	0
5	0.1	0.08	0.1	0
6	0.15	0.08	0.1	0
7	0.2	0.1	0.1	0.05
8	0.25	0.2	0.1	0.1
9	0.3	0.35	0.15	0.15
10	0.4	0.35	0.15	0.25
11	0.6	0.25	0.15	0.4
12	0.7	0.25	0.15	0.6
13	0.9	0.25	0.2	0.8
14	0.9	0.1	0.2	1
15	0.8	0.4	0.15	0.9
16	0.6	0.4	0.15	0.8
17	0.5	0.4	0.2	0.7
18	0.7	0.5	0.2	0.5
19	0.9	0.5	0.2	0.3
20	1	0.2	0.2	0.1
21	0.8	0.3	0.2	0.05
22	0.7	0.25	0.2	0
23	0.6	0.25	0.15	0
24	0.4	0.25	0.15	0

from the home side to the grid side. At hours 13–14 the home has excess energy and it is sent to the grid for making revenue. As well, at hour 19 the electricity from the grid is expensive and it is not economic to buy energy from the grid. Therefore, the home discharges electric vehicles for sending energy to the grid rather than taking energy from the grid. The load power is also scheduled and regulated with regard to

TABLE 3.3 Patterns of electric vehicles in parking station.

Hour	Electric vehicle number			
	1	2	3	4
1	1	0	0	0
2	1	0	0	0
3	1	1	0	0
4	1	1	0	0
5	1	1	0	0
6	1	1	0	0
7	1	1	0	0
8	1	1	0	0
9	1	1	0	0
10	0	1	1	0
11	0	1	1	0
12	0	1	1	0
13	0	0	1	0
14	0	0	1	1
15	0	0	1	1
16	0	0	1	1
17	0	0	1	1
18	0	0	1	1
19	0	0	1	1
20	0	0	1	1
21	0	0	1	1
22	0	0	1	1
23	0	0	0	1
24	0	0	0	1

the electricity price. It is seen that the loads are shifted to the initial hours of the day for consuming cheap electricity. In hours 1−6, the biggest parts of the loads are supplied. The parking station exchanges energy with the home properly. It is seen that the parking station injects power into the home for many hours. As well, the received power from home is scheduled to be at nonexpensive hours. The surplus of solar

TABLE 3.4 Optimized loading conditions for HEMS with parking station.

Hour	Fixed load (kW)	Interpretable load (kW)	Shiftable load (kW)	Curtailable load (kW)	Curtailed load (%)
1	4	15	10	1	0
2	3	15	10	1	0
3	3	15	10	1	0
4	2	15	0	0.8	0
5	2	15	0	0.8	0
6	3	15	0	0.8	0
7	4	0	10	1	0
8	5	0	0	2	0
9	6	0	0	3.5	0
10	8	0	0	3.5	0
11	12	0	0	2.5	0
12	14	0	0	2.5	0
13	18	0	0	2.5	0
14	18	0	0	1	0
15	16	0	0	4	0
16	12	0	0	4	0
17	10	0	0	4	0
18	14	0	0	5	0
19	18	0	0	5	0
20	20	0	0	2	0
21	16	0	0	3	0
22	14	0	0	2.5	0
23	12	0	0	2.5	0
24	8	0	0	2.5	0

energy is stored by the parking station as seen at hour 15. The diesel generator is not operated for many hours because the home is not faced a shortage in supplying energy.

The equilibrium of power must be passed at all hours of the day. The balance of power in the home at some typical hours is presented in Fig. 3.2. At hour 3, the grid and parking station produce 96% and 4% of

TABLE 3.5 Optimized powers of resources for HEMS with parking station.

Hour	Grid (kW)	Load (kW)	Parking station (kW)	Solar (kW)	Diesel generator (kW)
1	30	30	0	0	0
2	29	29	0	0	0
3	27.8	29	−1.2	0	0
4	-2.2	17.8	−20	0	0
5	17.8	17.8	0	0	0
6	50	18.8	31.2	0	0
7	19.75	15	5.5	0.75	0
8	50	7	44.5	1.5	0
9	−35.25	9.5	−42.5	2.25	0
10	7.75	11.5	0	3.75	0
11	38.5	14.5	30	6	0
12	50	16.5	42.5	9	0
13	−45	20.5	−48.5	12	5
14	−16	19	−20	15	0
15	50	20	48.5	13.5	5
16	50	16	50	12	4
17	3.5	14	0	10.5	0
18	6.5	19	0	7.5	5
19	−31.5	23	−50	4.5	0
20	20.5	22	0	1.5	0
21	13.25	19	0	0.75	5
22	11.5	16.5	0	0	5
23	50	14.5	35.5	0	0
24	25	10.5	14.5	0	0

power to supply the load. At hour 13, the solar system, parking station, and diesel generator produce power, and such power is consumed by the grid and loads. The grid receives power from the home and appears as a load that consumes electricity.

Fig. 3.3 shows the charging–discharging powers of the parking station and it is clear that the discharge capability of electric vehicles is

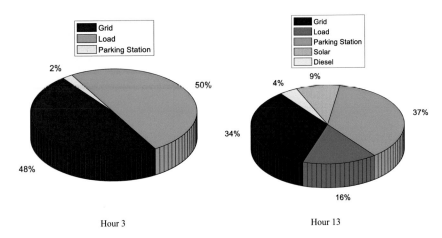

FIGURE 3.2   Balance of power in home at hours 3 and 13.

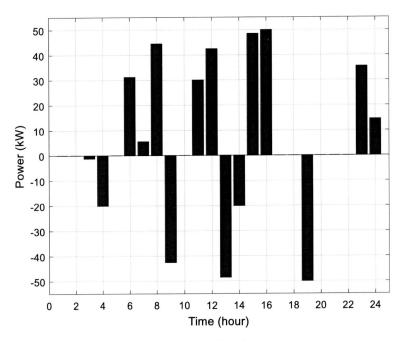

FIGURE 3.3   Charging–discharging powers of parking station.

properly used by HEMS. At many hours like 4, 9, 13, and 19 the parking station operates as a resource and produces energy for the home. The biggest discharge appears at hour 19 when the home is faced the high-priced electricity and HEMS tries to sell electricity to the external grid for making a profit. Apart from the scheduled discharging operation, the charging operation is also optimized to have the minimum energy cost. For instance, the charging process is not done at hours 17–22 when the electricity is expensive.

Fig. 3.4 shows the optimized powers of the grid and diesel generator. It is seen that the diesel generator is not operated too much and is only used at some limited hours. As well, the grid power changes from the positive direction (from the grid to the home) to the negative direction (from the home to the grid) and vice-versa for many hours.

There are four electric vehicles in the parking station. The power and energy of those electric vehicles are shown in Figs. 3.5 and 3.6. Electric vehicle 1 shows only one discharge operation and electric vehicle 2 shows 2 discharging operations. Both the electric vehicles are fully charged when they are leaving the station at hours 9 and 12, respectively. Electric vehicles 3 and 4 are also discharged at some hours like 13, 19, 14, and 21. They are also fully charged when leaving the station at hours 22 and 24, respectively. It is seen that the HEMS utilizes the electric vehicles properly to deal with energy costs at different hours but satisfies their operational constraints such as having full charge at the departure time.

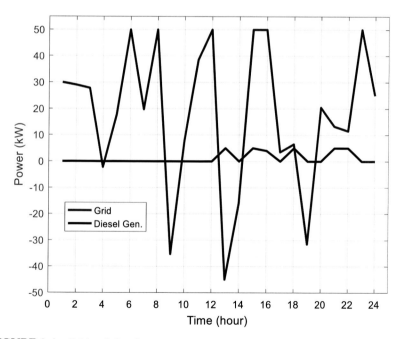

FIGURE 3.4   Grid and diesel generator powers.

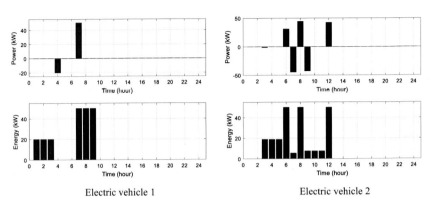

FIGURE 3.5   Power and energy of electric vehicles 1 and 2.

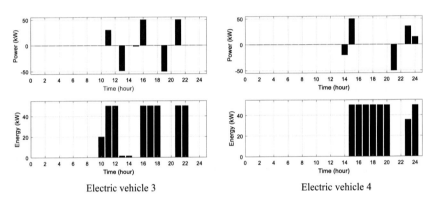

FIGURE 3.6   Power and energy of electric vehicles 3 and 4.

TABLE 3.6   Impacts of initial energy on energy cost.

Initial energy (kWh)	Daily energy cost ($/day)
0	66.96
10	61.46
20	55.96
30	50.46
40	44.96

Table 3.6 presents the impacts of initial energy inside the electric vehicle battery on the daily energy cost. It is seen that the higher initial energy in the battery of electric vehicles reduces the daily energy cost

because they can participate in the energy management system with more flexibility as well as they need a smaller amount of energy to be fully charged.

3.2.4 Parking station without discharge capability

Both the charging and discharging powers of electric vehicles are optimized and scheduled to minimize energy costs. The charging power is scheduled to be at the times when electricity is not expensive and the discharging power is scheduled to reduce the energy cost or deal with energy shortage. As a result, if the discharge capability is not utilized, the energy cost would be increased. This point is simulated and studied here. In order to model the electric vehicles without discharge capability, some relationships of the model are reformulated as Eqs. (3.21) and (3.22).

$$
\begin{aligned}
u_{cev}^{t,v} &\le 1 \\
u_{dev}^{t,v} &= 0 \\
\forall t &\in T \\
\forall v &= V
\end{aligned}
\tag{3.21}
$$

$$
\begin{aligned}
P_{dev}^{t,v} &= 0 \\
\forall t &\in T \\
\forall v &= V
\end{aligned}
\tag{3.22}
$$

Without discharge capability, the daily energy cost increases to 61.440 \$/day which shows about a 10% increase compared to the model with discharge capability. The electric vehicles are only charged as indicated by Fig. 3.7. It should be mentioned that the charging power of electric vehicles is still optimized and scheduled by HEMS. However, since the discharge capability is not used, the HEMS cannot utilize the electric vehicles as a generating resource in the energy management system and only is able to schedule their charging process. It is seen that the charging powers are scheduled at the hours when electricity is not expensive.

3.2.5 Limited capacity for line

In the HEMS, the shortage and surplus of energy are dealt with by the upstream grid. As a result, the capacity of the line between the home and the grid makes a direct impact on the HEMS. If the line capacity is limited, the home receives less power from the grid as well as it only can send some part of the energy surplus to the grid. In such situations, the resources need to be rescheduled. As a typical example, the capacity of the line is limited to 15 kW and the optimized powers of resources with

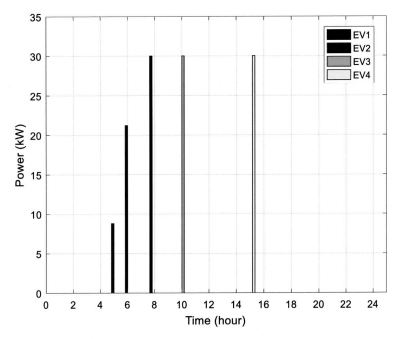

FIGURE 3.7   Charged power by each electric vehicle.

15 kW line capacity are listed in Table 3.7. It is seen that the exchanged power with the grid is positive at all hours which means the home cannot send energy to the grid and only receives energy from the grid. The discharge operation of the parking station is limited significantly. The diesel generator operation is increased considerably to deal with the shortage of power caused by limited line capacity. The diesel generator operation is shown in Fig. 3.8. Under 50 kW line capacity, the diesel generator operates for 6 hours but under 15 kW line capacity, the diesel generator operates with maximum capacity at almost all hours of the day.

3.2.6 Stochastic HEMS with parking station

In the stochastic HEMS with the parking station, some parameters are associated with uncertainty, and the uncertainty is modeled by a set of scenarios of performance. Some relationships of the developed model need to be reformulated. The objective function is re-expressed as Eq. (3.23). The expected value of electricity cost is calculated in the first term of the objective function. The other terms are the same as the deterministic model. The grid, solar, and load powers are assumed to

TABLE 3.7 Optimized powers of resources with 15 kW line capacity.

Hour	Grid (kW)	Load (kW)	Parking station (kW)	Solar (kW)	Diesel generator (kW)
1	15	20	0	0	5
2	15	19	1	0	5
3	15	19	1	0	5
4	15	17.8	2.2	0	5
5	15	17.8	2.2	0	5
6	15	18.8	0	0	3.8
7	15	5	15.75	0.75	5
8	15	7	14.5	1.5	5
9	15	9.5	12.75	2.25	5
10	15	11.5	12.25	3.75	5
11	15	14.5	11.5	6	5
12	15	16.5	12.5	9	5
13	7.85	20.5	4.35	12	5
14	15	19	11	15	0
15	15	30	0	13.5	1.5
16	15	26	1	12	0
17	13.5	24	0	10.5	0
18	15	27.5	0	7.5	5
19	15	24.5	0	4.5	5
20	15	22	−0.5	1.5	5
21	13.25	19	0	0.75	5
22	15	16.5	3.5	0	5
23	15	14.5	5.5	0	5
24	15	10.5	9.5	0	5

be associated with uncertainty and they are defined over a set of scenarios of performance as indicated by Eq. (3.24). The load power is expressed by Eq. (3.25) and the capacity of the line is rearranged as Eq. (3.26). The other relationships are the same as the deterministic model [12].

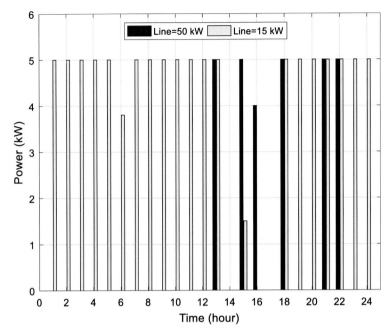

FIGURE 3.8 Diesel generator operation with 15 kW and 50 kW line capacities.

$$D_{ec} = \sum_{s=1}^{S} \left(\sum_{t=1}^{T} \left(P_{grid}^{t,s} \times E_p^t \right) \times O_{pr}^s \right) + \sum_{t=1}^{T} \left(P_{diesel}^t \times F_p^t \right) + \sum_{t=1}^{T} \left(1 - K_{cload}^t \right) \times C_{cload}^t$$

(3.23)

$$P_{grid}^{t,s} = P_{load}^{t,s} + P_{cp}^t - P_{dp}^t - P_{solar}^{t,s} - P_{diesel}^t$$
$$\forall t \in T$$
$$\forall s \in S$$

(3.24)

$$P_{load}^{t,s} = P_{fload}^{t,s} + P_{iload}^t + P_{sload}^t + \left(P_{cload}^t \times K_{cload}^t \right)$$
$$\forall t \in T$$
$$\forall s \in S$$

(3.25)

$$\left| P_{grid}^{t,s} \right| \leq P_{grid}^r$$
$$\forall t \in T$$
$$\forall s \in S$$

(3.26)

The solar and fixed load powers are defined over a set of scenarios. In this example, seven scenarios are selected to cover all areas of uncertainty. Fig. 3.9 shows the solar energy including uncertainty and it is seen that the scenarios properly cover all areas of uncertainty. The numerical values of the scenarios are listed in Table 3.8. The uncertainty area and scenarios of fixed load are presented in Fig. 3.10 and Table 3.9.

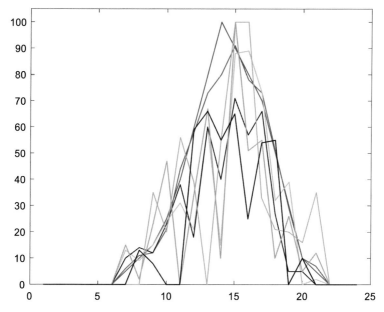

FIGURE 3.9 Schematic of solar energy alteration with uncertainty.

The defined stochastic HEMS with the parking station is simulated and the daily energy cost is minimized as listed in Table 3.10. It is seen that the uncertainty in the parameters increases the energy cost by about 11% because the HEMS has to utilize more resources. The HEMS changes the operating patterns of adjustable loads and resources from the nominal operating condition in order to deal with the uncertainty.

Table 3.11 lists the grid power under all scenarios of performance and it is seen that the grid power changes together with the alteration of parameters in the home and such a flexible grid power assists the home to deal with uncertainty in the resources and loads powers. Fig. 3.11 shows the total load of the home under all scenarios of performance and the variations of load power are obvious. Such load variation is dealt with by the upstream grid.

As it was stead, the uncertainty in the parameters changes the optimal operating condition of the system. Table 3.12 demonstrates the optimized operating condition of loads for both deterministic and stochastic models. It is clear that the stochastic model changes the operating hours of loads toward the hours when electricity is more expensive. Such a point increases the energy cost.

The power and energy of all the electric vehicles are shown in Fig. 3.12. The most important point is that electric vehicles are optimized to have a unique operating pattern under all scenarios of performance and their operation is not altered together with changing the

TABLE 3.8   Numerical values of solar energy in per-unit under each scenario.

	Scenario number						
Hour	1	2	3	4	5	6	7
1	0	0	0	0	0	0	0
2	0	0	0	0	0	0	0
3	0	0	0	0	0	0	0
4	0	0	0	0	0	0	0
5	0	0	0	0	0	0	0
6	0	0	0	0	0	0	0
7	0.05	0.06	0.13	0.1	0.15	0	0
8	0.1	0.11	0.09	0.14	0.02	0	0.13
9	0.15	0.12	0.15	0.12	0.23	0.35	0.08
10	0.25	0.21	0.24	0.23	0.47	0.2	0
11	0.4	0.44	0.31	0.38	0	0.56	0
12	0.6	0.58	0.21	0.18	0.34	0.39	0.59
13	0.8	0.73	0.6	0.6	0.67	0	0.66
14	1	0.8	0.15	0.4	0.1	0.55	0.55
15	0.9	0.91	0.88	0.71	1	1	0.65
16	0.8	0.78	0.89	0.57	0.51	1	0.25
17	0.7	0.73	0.73	0.66	0.55	0.33	0.54
18	0.5	0.51	0.32	0.27	0.1	0.21	0.55
19	0.3	0.29	0.39	0.05	0.26	0.2	0
20	0.1	0.1	0	0.05	0.05	0.16	0.1
21	0.05	0.07	0.02	0	0.12	0.35	0
22	0	0	0	0	0	0	0
23	0	0	0	0	0	0	0
24	0	0	0	0	0	0	0
Probability	0.2	0.2	0.15	0.15	0.1	0.1	0.1

parameters. Such a point increases the lifetime of their batteries and avoids too many charging–discharging cycles. This robust operating pattern which satisfies all scenarios of performance is also practical because it is not feasible to change the charging-discharging pattern of

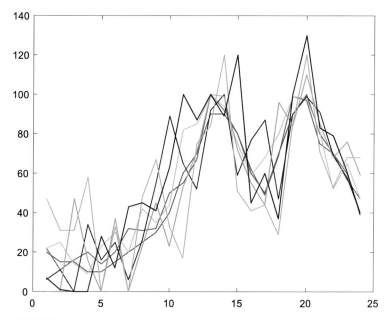

FIGURE 3.10 Schematic of fixed load power alteration with uncertainty.

electric vehicles alongside any change in solar energy or load power. All the electric vehicles are fully charged when they are leaving the station.

Under the stochastic model, the diesel generator shows more operation as depicted in Fig. 3.13. The uncertainty in the parameters like solar energy or load power causes the HEMS to utilize more local resources such as diesel generator. This point faces the home with more energy cost.

3.3 Residential microgrid with charging station

The previous home integrated with the parking station is extended to a residential microgrid integrated with the electric vehicle charging station. Fig. 3.14 shows the residential microgrid with renewable and nonrenewable resources, loads, external grid, and electric vehicle charging station [13]. The loads, solar system, and diesel generator are modeled, optimized, and operated the same as in the previous case study. The energy management system regulates the exchanged power with the grid the same as in the previous case study. The power can be received from or sent to the grid based on economic or technical dictates.

TABLE 3.9   Numerical values of load powers in per-unit under each scenario.

Hour	Scenario number						
	1	2	3	4	5	6	7
1	0.2	0.22	0.22	0.06	0	0.47	0.07
2	0.15	0.11	0.25	0.11	0	0.31	0.01
3	0.15	0.16	0.14	0	0.47	0.31	0
4	0.1	0.2	0.09	0.34	0.16	0.58	0
5	0.1	0.14	0.23	0.16	0	0	0.28
6	0.15	0.2	0.3	0.25	0.37	0.33	0.12
7	0.2	0.32	0.2	0.06	0	0	0.43
8	0.25	0.31	0.42	0.26	0.23	0.48	0.45
9	0.3	0.32	0.35	0.55	0.45	0.67	0.41
10	0.4	0.5	0.44	0.89	0.23	0.33	0.63
11	0.6	0.55	0.82	0.65	0.53	0.17	1
12	0.7	0.67	0.85	0.52	0.71	0.75	0.87
13	0.9	1	0.99	0.92	1	0.84	1
14	0.9	0.92	0.94	1	0.99	1.2	0.89
15	0.8	0.8	0.75	0.59	0.72	0.51	1.2
16	0.6	0.62	0.59	0.77	0.55	0.41	0.45
17	0.5	0.49	0.68	0.87	0.44	0.44	0.6
18	0.7	0.69	0.8	0.47	0.96	0.29	0.37
19	0.9	0.99	0.99	0.9	0.84	0.85	1
20	1	0.97	0.98	0.99	1.1	1.2	1.3
21	0.8	0.75	0.81	0.91	0.85	0.71	0.83
22	0.7	0.7	0.52	0.69	0.68	0.53	0.79
23	0.6	0.57	0.65	0.58	0.76	0.68	0.61
24	0.4	0.47	0.47	0.4	0.59	0.68	0.39
Probability	0.2	0.2	0.15	0.15	0.1	0.1	0.1

TABLE 3.10   Daily energy cost with and without uncertainty.

Model	Daily energy cost ($/day)
Deterministic model	55.965
Stochastic model	62.429

TABLE 3.11 Grid power in kW under all scenarios of performance.

Hour	Scenario number						
	1	2	3	4	5	6	7
1	35	35.4	35.4	32.2	31	40.4	32.4
2	6.4	5.6	8.4	5.6	3.4	9.6	3.6
3	19	19.2	18.8	16	25.4	22.2	16
4	40.4	42.4	40.2	45.2	41.6	50	38.4
5	46.4	47.2	49	47.6	44.4	44.4	50
6	28.8	29.8	31.8	30.8	33.2	32.4	28.2
7	29.25	31.5	28.05	25.7	23.75	26	34.6
8	16.9	17.95	20.45	16.5	17.7	23	20.45
9	3.6	4.45	4.6	9.05	5.4	8	6.85
10	7.75	10.35	8.7	17.85	1.05	7.1	16.1
11	8.5	6.9	14.25	9.8	13.1	−2.5	22.5
12	41.15	40.85	50	43.85	45.25	45.3	44.7
13	−46.5	−43.45	−41.7	−43.1	−42.55	−35.7	42.4
14	−2.75	0.65	10.8	8.25	12.55	10	3.8
15	38.25	38.1	37.55	36.9	35.15	30.95	50
16	43.15	43.85	41.6	50	46.5	36.35	48.4
17	−32.35	−33	−29.2	−24.35	−31.3	−28	27.95
18	6.5	6.15	11.2	5.35	17.7	2.65	-0.85
19	13.5	15.45	13.95	17.25	12.9	14	20
20	−34.5	−35.1	−33.4	−33.95	−31.75	−31.4	28.5
21	13.25	11.95	13.9	16.2	13.2	6.95	14.6
22	48.2	48.2	44.6	48	47.8	44.8	50
23	46.8	46.2	47.8	46.4	50	48.4	47
24	28.2	29.6	29.6	28.2	32	33.8	28

In the charging station, several electric vehicles come into the station at each time interval and they need to be fully charged and leave the station. Unlike the parking station, electric vehicles are not parked in the station for hours. The energy management system should optimize the charging process of each arrived electric vehicle based on the following rules;

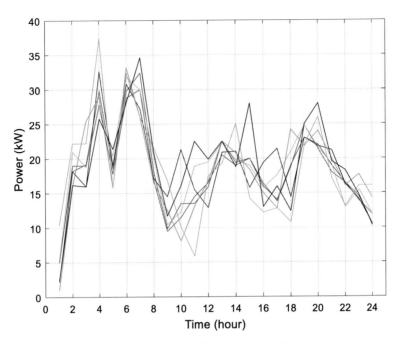

FIGURE 3.11   Total load of home under all scenarios of performance.

TABLE 3.12   Load energy management under deterministic and stochastic models.

Hour	Interpretable load (kW)		Shiftable load (kW)	
	Deterministic	Stochastic	Deterministic	Stochastic
1	15	0	10	0
2	15	15	10	0
3	15	15	10	0
4	15	15	0	10
5	15	15	0	0
6	15	15	0	10
7	0	15	10	10
8	0	0	0	10
9–24	0	0	0	0

- During the time when the electric vehicle is plug-in, the energy management system must schedule the charging process for having minimum energy cost. In general, the energy should be stored at off-peak hours when the electricity is inexpensive.

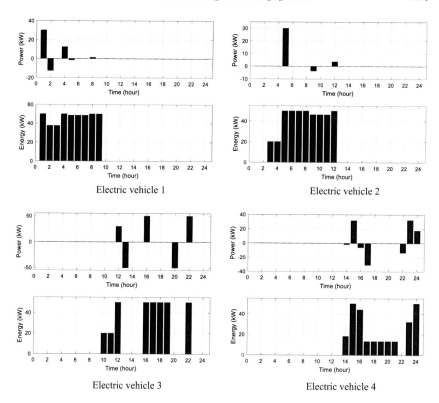

FIGURE 3.12 Power and energy of electric vehicles 1−4 under the stochastic model.

- The battery can also be discharged, when necessary, based on economic or technical conditions.
- The electric vehicle must be fully charged after a predefined time period. For instance, each electric vehicle needs to remain one hour in the station and after that time period, it must be fully charged and ready for leaving the station. During this one hour, the charging or discharging process is managed by the energy management system.

3.3.1 Input data and mathematical model

The mathematical model for a residential microgrid with the electric vehicle charging station is similar to the HEMS with the parking station except for the charging time of electric vehicles. Here, there is a pattern for inputting electric vehicles into the charging station and each arrival electric vehicle must be fully charged after a short duration of time. The 24-hour time-period of the day is modeled as 96 time-intervals each one 15 minutes [3,6]. It is assumed that the electric vehicles need to remain

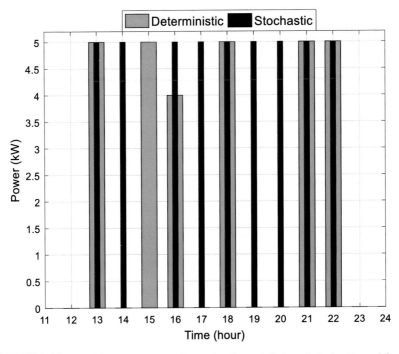

FIGURE 3.13　Diesel generator operation under deterministic and stochastic models.

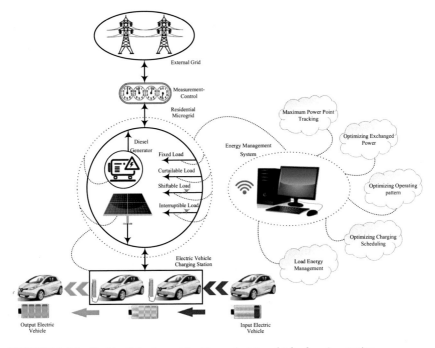

FIGURE 3.14　Residential microgrid with an electric vehicle charging station.

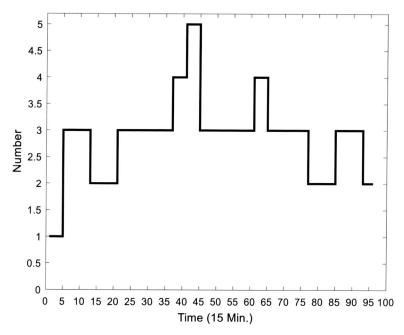

FIGURE 3.15   Number of electric vehicles entered to charging station at each time-interval.

in the charging station for two hours (8 time-intervals) to get fully charged. Fig. 3.15 presents the number of electric vehicles that arrive at the charging station at each time interval. Each electric vehicle is fully charged at 8 time-intervals and then leaves the station. The optimization programming formulation, objective function, and constraints are the same as HEMS with the parking station. The information of this test system such as solar energy, loads, and the diesel generator are selected similarly to the previous case study and are taken from that example. The electricity price over 96 time-intervals is depicted in Fig. 3.16 and solar energy is shown in Fig. 3.17. The capacity of the line between the microgrid and the external grid is 120 kW. The interpretable load needs to operate at 8 consecutive time-intervals.

3.3.2 Simulations and discussions

The optimization programming is solved to find the optimal operating condition of the residential microgrid and minimize the daily energy cost. The energy cost is minimized as 228.952 $/day. The optimal patterns of interpretable and shiftable loads are listed in Table 3.13. The interpretable load is operated at 8 consecutive time-intervals.

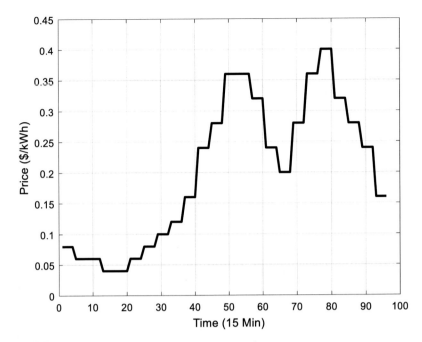

FIGURE 3.16   Electricity price over 96 time-intervals.

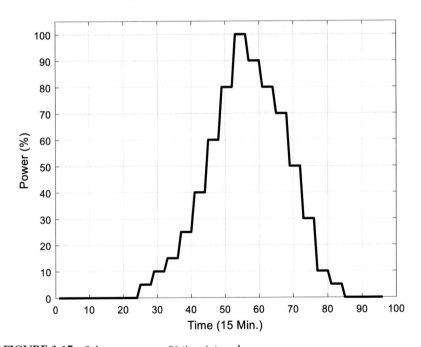

FIGURE 3.17   Solar energy over 96 time-intervals.

TABLE 3.13   Optimized loading conditions for residential microgrid with charging station.

Time interval	Interpretable load (kW)	Shiftable load (kW)
1–12	0	0
13	15	0
14	15	10
15	15	10
16	15	0
17	15	0
18	15	0
19	15	10
20	15	10
21–96	0	0

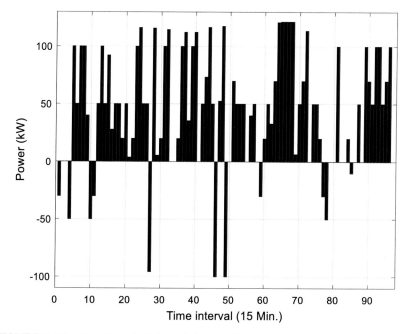

FIGURE 3.18   Charging scheduling of electric vehicle charging station.

Fig. 3.18 shows the charging scheduling of the electric vehicle charging station. It is seen that the charging station appears as a resource and produces electricity at many time intervals. The maximum produced power by the electric vehicle charging station is 100 kW at time intervals 47 and 49. The charged power to the charging station shows that the energy management system properly schedules the charged power and reduces the level of power when electricity is expensive.

The traded power with the grid and the produced power by the diesel generator are depicted in Fig. 3.19. The power to the grid becomes negative at many time intervals which demonstrates selling power to the grid by the microgrid. The diesel generator is also operated especially at on-peak loading time intervals.

Table 3.14 presents the optimized charging scheduling of electric vehicles 1 and 4. The electric vehicle 1 is discharged at two time-intervals and charged at many other time periods. Finally, this vehicle is fully charged at time-interval 50 and leaves the station. Electric vehicle 4 also participates in the energy management problem and shows discharge operation at time intervals 9 and 12. After discharge operation at time-interval 12, the battery of this electric vehicle is completely depleted but it is charged again at time-intervals 13–16 and gets fully charged at time-interval 16 in order to leave the station. It is clear that the energy management problem optimally utilizes electric vehicles taking into account their operational constraints such as being fully charged at proper time intervals.

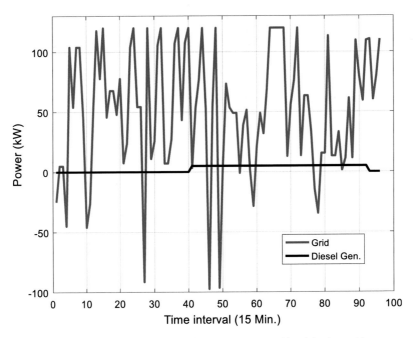

FIGURE 3.19   Powers of grid and diesel generator in a residential microgrid.

TABLE 3.14   Charging scheduling of electric vehicles 1 and 4.

Hour	Electric vehicle 1			Electric vehicle 4		
	Charged power (kW)	Discharged power (kW)	Energy (kWh)	Charged power (kW)	Discharged power (kW)	Energy (kWh)
1	0	30	12.5	0	0	0
2	0	0	12.5	0	0	0
3	0	0	12.5	0	0	0
4	0	50	0	0	0	0
5	50	0	12.5	0	0	0
6	50	0	25	0	0	0
7	50	0	37.5	0	0	0
8	50	0	50	0	0	0
9	0	0	0	0	30	12.5
10	0	0	0	0	0	12.5
11	0	0	0	0	0	12.5
12	0	0	0	0	50	0
13	0	0	0	50	0	12.5
14	0	0	0	50	0	25
15	0	0	0	50	0	37.5
16	0	0	0	50	0	50
17–96	0	0	0	0	0	0

The charging scheduling of electric vehicles 12 and 31 is demonstrated in Table 3.15. Electric vehicle 12 is discharged two times and electric vehicle 31 shows 3 times discharge operations. They get fully charged at time intervals 40 and 88, respectively.

3.3.3 Charging station without discharge capability

The proposed energy management system optimizes the energy cost by performing two strategies as follows:

* Optimizing the charged power to the electric vehicles for achieving minimum energy cost and doing the charging process at nonexpensive hours.

TABLE 3.15   Charging scheduling of electric vehicles 12 and 31.

Hour	Electric vehicle 12			Electric vehicle 31		
	Charged power (kW)	Discharged power (kW)	Energy (kWh)	Charged power (kW)	Discharged power (kW)	Energy (kWh)
1–32	0	0	0	0	0	0
33	0	0	20	0	0	0
34	50	0	32.5	0	0	0
35	20	0	37.5	0	0	0
36	50	0	50	0	0	0
37	0	37.75	40.563	0	0	0
38	0	12.25	37.5	0	0	0
39	0	0	37.5	0	0	0
40	50	0	50	0	0	0
41–80	0	0	0	0	0	0
81	0	0	0	50	0	32.5
82	0	0	0	0	50	20
83	0	0	0	0	50	7.5
84	0	0	0	0	30	0
85	0	0	0	50	0	12.5
86	0	0	0	50	0	25
87	0	0	0	50	0	37.5
88	0	0	0	50	0	50
89–96	0	0	0	0	0	0

- Optimizing the discharged power from the electric vehicles in order to attain maximum profit from selling the discharged energy to the grid.

When the discharge capability is not allowed, the second item is not realized and it probably increases the energy cost. The charging scheduling of electric vehicles 1 and 4 without discharge capability is presented in Table 3.16. It is clear that the system only optimizes the charged power to the electric vehicles and all of them are fully charged at the departure time. As stated, disallowing the discharging process increases the daily energy cost by about 5% as presented in Table 3.17.

TABLE 3.16 Charging scheduling of electric vehicles 1 and 4 without discharge capability.

	Electric vehicle 1			Electric vehicle 4		
Hour	Charged power (kW)	Discharged power (kW)	Energy (kWh)	Charged power (kW)	Discharged power (kW)	Energy (kWh)
1	0	0	20	0	0	0
2	0	0	20	0	0	0
3	0	0	20	0	0	0
4	0	0	20	0	0	0
5	50	0	32.5	0	0	0
6	20	0	37.5	0	0	0
7	0	0	37.5	0	0	0
8	50	0	50	0	0	0
9	0	0	0	0	0	20
10	0	0	0	0	0	20
11	0	0	0	0	0	20
12	0	0	0	0	0	20
13	0	0	0	50	0	32.5
14	0	0	0	50	0	45
15	0	0	0	0	0	45
16	0	0	0	20	0	50
17–96	0	0	20	0	0	0

TABLE 3.17 Daily energy cost with and without discharge capability.

Case	Daily energy cost ($/day)
With discharge capability	228.952
Without discharge capability	240.612

The discharge capability not only makes positive impacts on the energy cost but also increases the resilience of the microgrid following events. Table 3.18 indicates the operation of the microgrid following the outage of the line at time intervals 77–80. When the line is disconnected,

TABLE 3.18 Operation of microgrid following outage of line at time intervals 77–80.

Operation	Daily energy cost ($/day)	
With discharge capability	Feasible	229.42
Without discharge capability	Infeasible	—

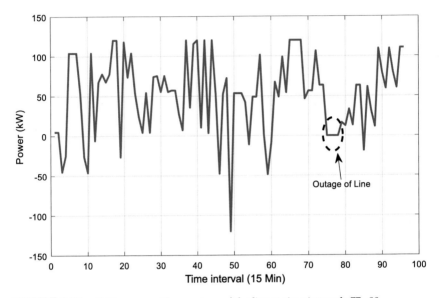

FIGURE 3.20   Grid power with an outage of the line at time intervals 77–80.

the system without discharge capability has infeasible operation because the local resources cannot handle the load demand. On the other hand, the system with discharge capability has a feasible operation with a daily energy cost equal to 229.42 $/day. The exchanged power with the grid following the outage of the line at time intervals 77–80 is depicted in Fig. 3.20. It is seen that the power at time intervals 77–80 is zero. The results demonstrate that the discharge capability of electric vehicles in the charging station improves the system resilience and assists the microgrid to supply the load demand under events and outages.

Table 3.19 lists the power of resources and loads in the microgrid following the outage of the line at time intervals 77–80. It is seen that the grid power is zero and the local resources including the charging station, solar system, and diesel generator supply the loads.

TABLE 3.19 Power balance in microgrid following an outage of the line at time intervals 77−80.

Time-interval	Grid (kW)	Load (kW)	Charging station (kW)	Solar (kW)	Diesel generator (kW)
77	0	23	13.5	4.5	5
78	0	23	13.5	4.5	5
79	0	22	15.5	1.5	5
80	0	22	15.5	1.5	5

3.3.4 Stochastic model for microgrid with charging station

Similar to the stochastic HEMS with the parking station, the grid, solar, and load powers are assumed to be associated with uncertainty and they are defined over a set of scenarios of performance [13]. The mathematical formulation, objectives, and constraints are similar to the stochastic HEMS with the parking station. In the charging station, the entered electric vehicles to the station need to be charged in a shorter time compared to the parking station, where the electric vehicles remain in the station for several hours. Additionally, in the charging station, there are more electric vehicles that enter the station and get fully charged. As a result, a more complicated charging scheduling is required.

In the introduced residential microgrid with the charging station, the load and solar energies are assumed as uncertain parameters, and a set of scenarios of performance are defined to model their uncertainty as depicted by Figs. 3.21 and 3.22. It is obvious that the scenarios of performance properly cover the uncertain area and all possible operating conditions are covered by the defined scenarios.

One of the most important points regarding optimization programming under uncertainty is to assign one of the resources to handle the uncertainty and alterations in the parameters. In this example, the uncertainty and alterations of load and solar energy are handled by the upstream grid. In other words, when solar energy increases under one scenario of performance, the excess of solar energy is absorbed by the upstream grid and when solar energy decreases under another scenario of performance, the shortage of solar energy is compensated by the external grid. As a result, the shortage or excess energy in the microgrid is handled by the grid. If the external grid is faced with an outage and the microgrid operates as islanded, the energy management system becomes infeasible because there is no resource to handle the uncertainty. In such situations, a curtailable load can be modeled over scenarios of performance to deal with uncertainty.

The other point regarding the optimization programming under uncertainty is to model some resources without uncertainty. In other words,

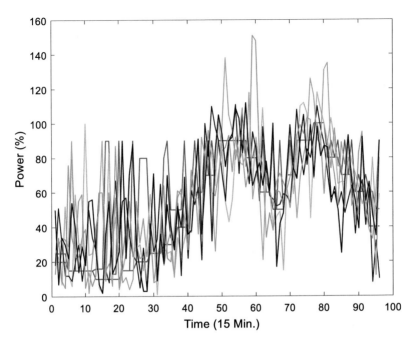

FIGURE 3.21 Scenarios of performance to model load energy uncertainty.

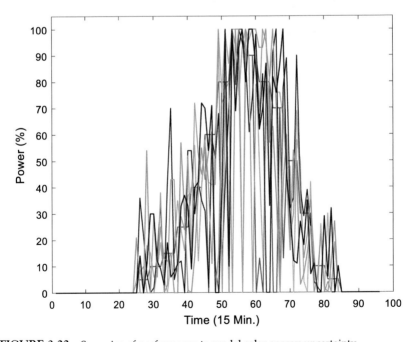

FIGURE 3.22 Scenarios of performance to model solar energy uncertainty.

some resources cannot change or must not change their daily operation together with any change in the solar or load energies. For instance, the charging scheduling of electric vehicles in the charging station should not be affected by scenarios of performance. As a result, the variables and parameters related to the electric vehicle charging station and electric vehicles are modeled excluding scenarios of performance. In such modeling, the electric vehicle charging station and electric vehicles have a unique operation pattern for 24 hours under all scenarios of performance.

In order to demonstrate the effects of uncertainty on the model, the results with and without uncertainty are presented in Table 3.20. It is seen that the stochastic model which comprises the scenarios of performance increases the daily energy cost by about 3.5% since the stochastic model needs to deal with alterations of loads and solar energy.

Fig. 3.23 demonstrates the exchanged power with the external grid under all scenarios of performance. The uncertainty introduced by scenarios changes the traded power and this power ranges from a minimum level to a maximum level. There are about 20 kW variations between the minimum and the maximum levels which is about 20% of the nominal level. Such alterations are required to handle the uncertainty in the resources and loads.

Figs. 3.24−3.26 demonstrate the energy of some typical electric vehicles in the charging station. Those electric vehicles arrive at the charging

TABLE 3.20 Daily energy cost with deterministic and stochastic models.

	Daily energy cost ($/day)
Deterministic model	228.952
Stochastic model	236.23

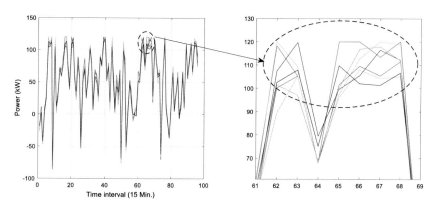

FIGURE 3.23   Exchanged power with the grid under all scenarios of performance.

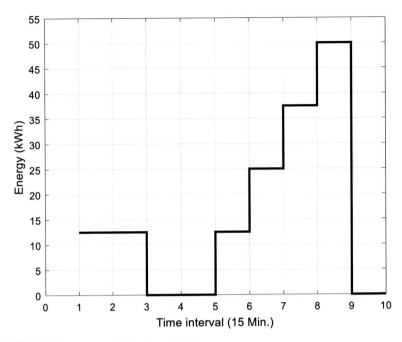

FIGURE 3.24   Energy of electric vehicle 1.

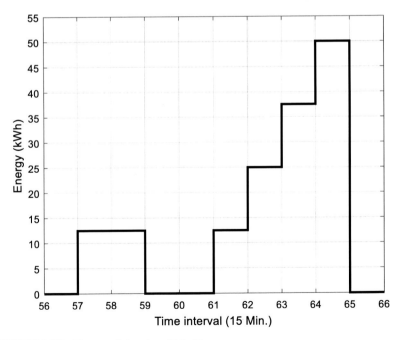

FIGURE 3.25   Energy of electric vehicle 12.

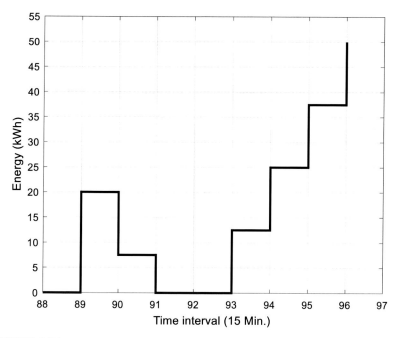

FIGURE 3.26 Energy of electric vehicle 34.

station and are connected to the electricity. Their charging process is optimized and scheduled and they are also discharged at some hours. Eventually, they get fully charged when leaving the charging station. For instance, electric vehicle 1 has 12.5 kWh of energy when comes to the charging station. It is discharged at time interval 3 and its battery is depleted completely. Afterward, it gets charged at time intervals 5–8 and it becomes fully charged when leaving the station. The electric vehicles 12 and 34 are also discharged completely at some time intervals and are then charged again to get fully charged. Such optimal charging-discharging is done for all the electric vehicles that arrive at the charging station during 24 hours.

3.4 Microgrid with battery swapping station

The operation of the battery swapping station is depicted in Fig. 3.27. In this station, the dead or dying batteries of electric vehicles are swapped for fully charged batteries in a matter of minutes. The depleted batteries remain in the station and are charged for the next vehicles. The batteries which are under charge may be used in the

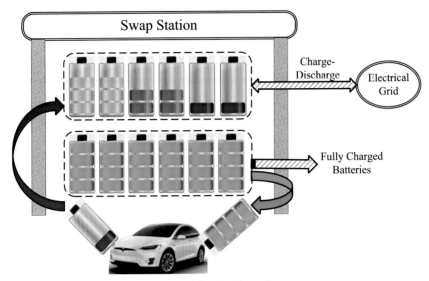

Exchanging Depleted Battery With Fully Charged
Battery

FIGURE 3.27   Operation of battery swapping station.

energy management system, where their charging is scheduled or they are discharged for injecting power into the system [14].

Fig. 3.28 shows the test microgrid with renewable and nonrenewable resources, loads, external grid, and battery swapping station. The loads and resources are modeled, optimized, and operated the same as in the previous case studies. The exchanged power with the grid is also modeled and managed the same as in the previous case study.

In the battery swapping station, the operation is based on the following procedure:

- Each input electric vehicle is equipped with one fully charged battery and its dead battery is added to the station.
- The batteries inside the swapping station are connected to the microgrid for charging scheduling. The charging power and the discharging power of these batteries are managed.
- At each time interval, the swapping station must have enough fully charged batteries to support the entered electric vehicles. If "n" electric vehicles arrive at the station at time interval "t", the energy management system charges "n" electric vehicles at time interval "$t-1$" and keeps them fully charged for swapping at time interval

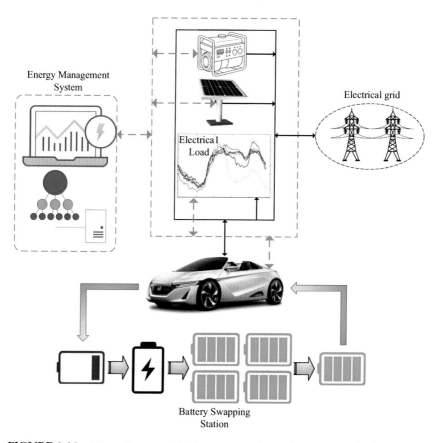

FIGURE 3.28 Microgrid connected to battery swapping station, resources, loads and grid.

"t". These electric vehicles are not discharged or involved in the energy management process because they should be kept fully charged for swapping at the next time interval.
- During swapping time, the power and energy of the swapped battery and the new dead battery in the station are set at zero because they are not connected to the electricity.
- Number of batteries inside the station must be greater than the maximum number of electric vehicles which arrive at the station at a specific time interval. Because when the depleted battery is separated from the electric car and added to the station, it cannot be immediately charged and get ready for the coming electric vehicle at the next time-period. As a result, the swapping station must have several batteries more than the maximum number of electric vehicles.

3.4.1 Mathematical model for battery swapping station

In the microgrid with the battery swapping station, the objective function is similar to the previous case studies. The exchanged power with the grid is modeled by Eq. (3.27). The total charged power and total discharged power of the battery swapping station are modeled by Eqs. (3.28) and (3.29). These total powers are calculated by adding the power of all batteries in the battery swapping station.

$$P^t_{grid} = P^t_{load} + P^t_{cbss} - P^t_{dbss} - P^t_{solar} - P^t_{diesel} \tag{3.27}$$
$$\forall t \in T$$

$$P^t_{cbss} = \sum_{b=1}^{B} \left(P^{t,b}_{cbs} \right) \tag{3.28}$$
$$\forall t \in T$$

$$P^t_{dbss} = \sum_{b=1}^{B} \left(P^{t,b}_{dbs} \right) \tag{3.29}$$
$$\forall t \in T$$

Every battery in the battery swapping station is charged or discharged like a regular battery as expressed by Eqs. (3.30)–(3.32). It is clear that the battery can either work in the charging state or discharging state at each specified time period. The defined binary variable by Eq. (3.30) is to confirm such an operation [15].

$$u^{t,b}_{cbs} + u^{t,b}_{dbs} \leq 1 \tag{3.30}$$
$$\forall t \in T$$
$$\forall b \in B$$

$$P^{t,b}_{cbs} \leq P^{r,b}_{bs} \times u^{t,b}_{cbs} \tag{3.31}$$
$$\forall t \in T$$
$$\forall b \in B$$

$$P^{t,b}_{dbs} \leq P^{r,b}_{bs} \times u^{t,b}_{dbs} \tag{3.32}$$
$$\forall t \in T$$
$$\forall b \in B$$

The stored energy of every battery inside the battery swapping station is calculated by Eq. (3.33) and the efficiency and rated capacity of each battery is denoted by Eqs. (3.34) and (3.35), respectively.

$$E^{t,b}_{bs} = E^{t-1,b}_{bs} + \left\{ P^{t,b}_{cbs} - \frac{P^{t,b}_{dbs}}{\eta_{bs}} \right\} \times I^t_{dur} \tag{3.33}$$
$$\forall t \in T$$
$$\forall b \in B$$

$$\eta_{bs} \frac{\sum_{t=1}^{T} \left(P_{dbs}^{t,b} \times I_{dur}^{t} \right)}{\sum_{t=1}^{T} \left(P_{cbs}^{t,b} \times I_{dur}^{t} \right)} \tag{3.34}$$

$$E_{bs}^{t,b} \leq E_{bs}^{r} \\ \forall t \in T \\ \forall b \in B \tag{3.35}$$

The battery swapping operation is modeled by Eqs. (3.36) and (3.37). In the battery swapping operation, the fully charged battery in the station is replaced with a depleted battery of an electric vehicle which arrives at the station. At the time of battery swapping, the fully charged battery is replaced with an empty battery. As well, at this time period, it is not possible to charge or discharge the battery because the swapping operation is taking place. As a result, at the swapping time- period, the energy and power of the battery are set at zero as modeled by Eq. (3.36). At the time period before swapping time, the battery needs to be fully charged for a replacement at the next time interval. As a result, the energy management system must arrange an adequate number of fully charged batteries at each time period for swapping at the next time period. This point is expressed by Eq. (3.37).

$$\begin{cases} E_{bs}^{t,b} = 0 \\ P_{cbs}^{t,b} = 0 \ \forall t = t_{swap} \forall b \in B \\ P_{dbs}^{t,b} = 0 \end{cases} \tag{3.36}$$

$$E_{bs}^{t,b} = E_{bs}^{r} \\ \forall t = \left(t_{swap} - 1 \right) \\ \forall b \in B \tag{3.37}$$

3.4.2 Data of microgrid with battery swapping station

In the proposed test system, 10 batteries are equipped in the battery swapping station. The rated capacity of the electric vehicle battery is 100 kWh, the rated power of the electric vehicle charger is 100 kW and the initial energy of batteries inside the battery swapping station is 100 kWh. The solar and load data are the same as in previous examples. Fig. 3.29 shows the number of electric vehicles that arrive at the battery swapping station at each hour. The capacity of the line between the external grid and the microgrid is set at 120 kW.

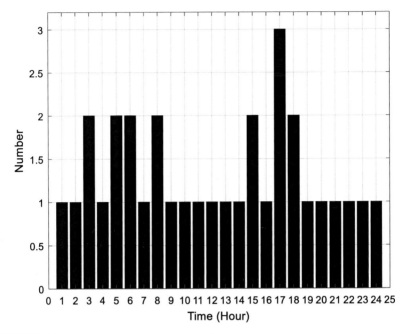

FIGURE 3.29 Number of electric vehicles that arrive at the battery swapping station at each hour.

3.4.3 Discussions on battery swapping operation

The proposed energy management system is carried out on the system with battery swapping operation and the daily energy cost is achieved equal to 375.955 $/day. Table 3.21 lists the optimized loading conditions for the microgrid with the battery swapping station. The adjustable loads such as interpretable and shiftable loads are scheduled to be operated at the hours with low-cost electricity. The load curtailment option is not employed by the energy management system.

Table 3.22 presents the optimized powers for the microgrid with the battery swapping station. The capacity of the line between the grid and the microgrid is 120 kW and this capacity is fully occupied during most of the hours. The produced power by the solar cell and diesel generator is totally consumed by the battery swapping station and loads. There is no excess energy to be injected into the upstream grid. As a result, the exchanged power with the external grid is always positive and the power is not transferred from the microgrid to the external grid. The charged power to the battery swapping station and the consumed power by loads are regulated and scheduled to deal with line capacity limitation. Since the capacity of the line is limited to 120 kW, the load and battery charging powers need to be shifted for satisfying such a

TABLE 3.21 Optimized loading conditions for microgrid with battery swapping station.

Hour	Fixed load (kW)	Interpretable load (kW)	Shiftable load (kW)	Curtailable load (kW)	Curtailed load (%)
1	4	15	10	1	0
2	3	15	1	1	0
3	3	15	1	1	0
4	2	15	0	0.8	0
5	2	15	0	0.8	0
6	3	15	0	0.8	0
7	4	0	0	1	0
8	5	0	0	2	0
9	6	0	0	3.5	0
10	8	0	0	3.5	0
11	12	0	0	2.5	0
12	14	0	0	2.5	0
13	18	0	0	2.5	0
14	18	0	0	1	0
15	16	0	0	4	0
16	12	0	0	4	0
17	10	0	0	4	0
18	14	0	0	5	0
19	18	0	0	5	0
20	20	0	0	2	0
21	16	0	10	3	0
22	14	0	0	2.5	0
23	12	0	8	2.5	0
24	8	0	10	2.5	0

constraint. These purposes are achieved by the energy management system where the loads are shifted, interrupted, or curtailed as well as the charging time of batteries is shifted and scheduled.

It should be noted that the power of the swapping station is always positive which may be assumed as no discharge operation in the

TABLE 3.22 Optimized powers for microgrid with battery swapping station.

Hour	Grid (kW)	Load (kW)	Swapping station (kW)	Solar (kW)	Diesel generator (kW)
1	30	30	0	0	0
2	120	20	100	0	0
3	120	20	100	0	0
4	120	17.8	102.2	0	0
5	120	17.8	102.2	0	0
6	120	18.8	101.2	0	0
7	120	5	115.75	0.75	0
8	120	7	114.5	1.5	0
9	120	9.5	117.75	2.25	5
10	120	11.5	112.25	3.75	0
11	120	14.5	116.5	6	5
12	120	16.5	117.5	9	5
13	103.5	20.5	100	12	5
14	120	19	121	15	5
15	120	20	118.5	13.5	5
16	120	16	121	12	5
17	120	14	121.5	10.5	5
18	27.9	19	21.4	7.5	5
19	118.5	23	100	4.5	0
20	115.5	22	100	1.5	5
21	120	29	96.75	0.75	5
22	11.5	16.5	0	0	5
23	120	22.5	100	0	2.5
24	20.5	20.5	0	0	0

swapping station. But at each hour, some batteries in the swapping station are discharged and some other ones are charged. The total power of the swapping station at each hour presents the net power which is total charged power minus total discharged power. If this value is always positive, it does not mean that there is no discharge operation on the batteries but it means that the total charged power to some batteries is greater than the total discharged power from the rest of the batteries. Fig. 3.30 shows the total discharged power from the swapping station which is the sum of discharged power by all batteries.

The power equilibrium at hours 4 and 18 is depicted in Fig. 3.31. At hour 4, the grid supplies the loads and battery swapping station. Since

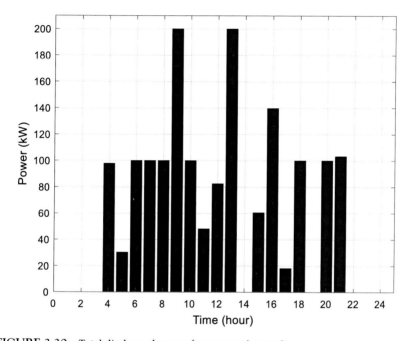

FIGURE 3.30   Total discharged power from swapping station.

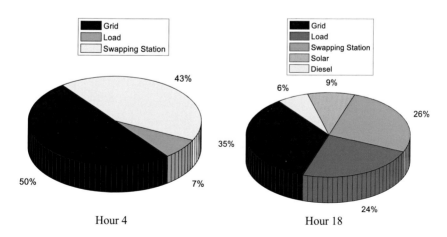

FIGURE 3.31   Equilibrium of power at hours 4 and 18.

the line capacity is enough to deal with those two consumers, the diesel generator is not activated. Solar energy is also zero at this time period. At hour 18, the grid, solar, and diesel generator produce 70%, 18%, and 12% of the required power of consumers (i.e., loads and battery swapping station). When the line capacity is not enough to supply the loads, the diesel generator is operated.

The swapping process is demonstrated in Table 3.23. The number of entered electric vehicles (EVs) at the station at each hour is presented

TABLE 3.23 Energy of batteries per kWh in swapping station.

Hour	Number of input EVs	Battery number									
		1	2	3	4	5	6	7	8	9	10
0	0	100	100	100	100	100	100	100	100	100	100
1	1	0	100	100	100	100	100	100	100	100	100
2	1	100	0	100	100	100	100	100	100	100	100
3	2	100	100	0	0	100	100	100	100	100	100
4	1	66.35	100	100	100	0	100	100	100	100	35.85
5	2	100	69.75	100	100	98.8	0	0	100	100	35.85
6	2	100	69.75	100	0	100	100	100	0	0	35.85
7	1	0	100	100	100	0	100	100	85.5	0	35.85
8	2	100	0	0	100	0	100	0	100	100	35.85
9	1	100	82.35	100	0	100	0	35.4	100	0	35.85
10	1	100	82.35	0	100	0	100	35.4	100	0	48.1
11	1	100	82.35	0	100	100	0	100	100	0	0
12	1	100	0	100	99.85	100	100	0	100	0	0
13	1	0	100	100	99.85	100	0	0	0	100	100
14	1	100	100	100	99.85	100	0	0	21	0	100
15	2	0	0	100	99.85	39.5	0	0	100	100	100
16	1	0	0	0	100	100	100	100	60.35	100	0
17	3	0	0	100	0	0	0	100	100	81.85	0
18	2	100	0	0	0	0	3.25	0	0	100	0
19	1	100	0	0	0	0	3.25	100	0	0	0
20	1	0	100	0	0	100	3.25	0	0	0	0
21	1	0	0	100	100	0	0	0	0	0	0
22	1	0	0	0	100	0	0	0	0	0	0
23	1	0	0	0	0	100	0	0	0	0	0
24	1	0	0	0	0	0	0	0	0	0	0

and the swapped batteries are highlighted. The energy at hour zero shows the initial energy inside the batteries. It is assumed that all the equipped batteries in the battery swapping station are fully charged at the beginning of the programming. At hour 1, only one electric vehicle comes to the station, and its battery is replaced with battery number 1. As a result, the energy of battery number 1 becomes zero because it is replaced with a depleted battery. Battery number 1 is swapped again at hours 7, 15, and 20. The energy inside this battery before the swapping hours is 100 kWh which shows the fully charged condition. In other words, the energy management system charges some batteries to the full charge level at each hour based on the pattern of entered electric vehicles at the next hour. If "n" electric vehicles arrive at the station at time-interval "t", the energy management system charges "n" electric vehicles to the full charge level at time interval "$t-1$" and keeps them fully charged for swapping at the next time interval. These electric vehicles are not discharged or used in the energy management process because they must be fully charged for swapping at the next time interval. The results demonstrate such operation, for instance, three electric vehicles come into the station at hour 17 and their batteries are switched by batteries 4, 5, and 6, where, these three batteries are fully charged at hour 16. This point is seen for all the swapped batteries. Apart from the swapping hours, the batteries are charged and discharged in the rest of the hours to participate in the energy management system. Each battery may be swapped several times within day hours.

Table 3.24 presents the charging scheduling of some batteries in the swapping station. It is clear that the batteries are charged and discharged at different hours of the day while they are fully charged right before the swapping hours. As well, the charged-discharged powers and energy are zero at the swapping hours. All the batteries are fully charged at hour zero which shows the initial energy of the batteries. This point is important because, at the first hour of the day, the batteries need to be fully charged to deal with the input electric vehicles at hour 1. All the batteries cannot be charged at hour 1 and be equipped on the input electric vehicles at the same time. Therefore, the possible solution is to use batteries with initial energy equal to full charge.

Fig. 3.32 shows the exchanged power with the grid as well as the power of the swapping station together with the electricity price. The charging scheduling in the battery swapping station properly assists the microgrid to reduce the exchanged power with the grid when electricity is expensive during hours like 13, 18, and 22. The received power from the grid is managed by the energy management system to be on the minimum level when electricity is expensive. Such a management is achieved by regulating load power or scheduling the charging time of batteries in the swapping station.

TABLE 3.24 Charging scheduling of batteries in the swapping station.

Hour	Battery 6				Battery 1			
	Charge (kW)	Discharge (kW)	Swapping status	Energy (kWh)	Charge (kW)	Discharge (kW)	Swapping status	Energy (kWh)
1	0	0	0	100	0	0	1	0
2	0	0	0	100	100	0	0	100
3	0	0	0	100	0	0	0	100
4	0	0	0	100	0	33.65	0	66.35
5	0	0	1	0	33.65	0	0	100
6	100	0	0	100	0	0	0	100
7	0	0	0	100	0	0	1	0
8	0	0	0	100	100	0	0	100
9	0	100	0	0	0	0	0	100
10	100	0	0	100	0	0	0	100
11	0	0	1	0	0	0	0	100
12	100	0	0	100	0	0	0	100
13	0	100	0	0	0	100	0	0
14	0	0	0	0	100	0	0	100
15	0	0	0	0	0	0	1	0
16	100	0	0	100	0	0	0	0

17	0	0	1	0	0	0	0	0
18	3.25	0	0	3.25	100	0	0	100
19	0	0	0	3.25	0	0	0	100
20	0	0	0	3.25	0	0	1	0
21	0	3.25	0	0	0	0	0	0
22	0	0	0	0	0	0	0	0
23	0	0	0	0	0	0	0	0
24	0	0	0	0	0	0	0	0

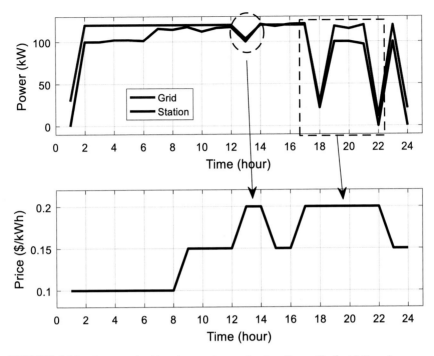

FIGURE 3.32   Powers of grid and swapping station together with electricity price.

TABLE 3.25   Daily energy cost with and without discharging capability in battery swapping station.

	Daily energy cost ($/day)
With discharge capability	375.955
Without discharge capability	395.945

3.4.4 Battery swapping without discharge capability

When the discharge capability is not used, the batteries in the swapping station can only be charged and cannot properly participate in the energy management system. This point is demonstrated in Table 3.25. Eliminating the discharge capability increases the daily energy cost by about 20 $/day or 5%. The discharge capability not only reduces the energy cost but also is able to support the system during the outages of elements. For instance, Table 3.26 presents the operation of the microgrid with a battery swapping station following the outage of the line at hours 19–20. When the line between the microgrid and the external grid is faced an outage, it means that the external grid is not available

TABLE 3.26 Operation of microgrid with battery swapping station following outage of line at hours 19−20.

	Operation	Daily energy cost ($/day)
With discharge capability	Feasible	395.945
Without discharge capability	Infeasible	—

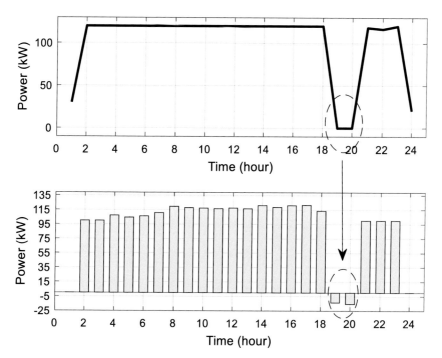

FIGURE 3.33 Grid and battery swapping station powers following an outage of the line at hours 19 and 20.

or the grid blackout happens. In such a situation, the microgrid can only be supplied by its own resources like the solar system, diesel generator, or battery swapping station. If these resources are not sufficient to supply the loads, the load curtailment option may be used by the energy management system. If all of those choices are not adequate, the operation becomes infeasible. As indicated by Table 3.26, the discharging capability assists the system in having feasible operation while in the system without discharging capability the available resources are not able to handle the load demand and the operation is infeasible.

Fig. 3.33 shows the exchanged power with the grid and the power of the battery swapping station following the outage of the line at hours 19 and 20.

It is clear that the battery swapping station acts as an energy resource and injects power to the system during the outage. The battery swapping station takes power from the microgrid at all hours except the outage hours.

Table 3.27 lists the optimized powers of the microgrid with battery swapping station following the outage of the line at hours 6 and 7. In these two hours, the grid power is zero and the load is supplied by the

TABLE 3.27 Optimized powers following an outage of the line at hours 6 and 7.

Hour	Grid (kW)	Load (kW)	Swapping station (kW)	Solar (kW)	Diesel generator (kW)
1	3	15	−12	0	0
2	120	14	106	0	0
3	120	14	106	0	0
4	120	2.8	122.2	0	5
5	120	2.8	122.2	0	5
6	0	3.8	1.2	0	5
7	0	5	0.75	0.75	5
8	120	7	119.5	1.5	5
9	120	9.5	117.75	2.25	5
10	120	11.5	117.25	3.75	5
11	120	14.5	116.5	6	5
12	120	16.5	117.5	9	5
13	120	20.5	116.5	12	5
14	120	19	121	15	5
15	120	20	118.5	13.5	5
16	120	16	121	12	5
17	120	14	121.5	10.5	5
18	120	19	113.5	7.5	5
19	120	38	91.5	4.5	5
20	120	37	89.5	1.5	5
21	120	34	91.75	0.75	5
22	120	31.5	89.9	0	1.4
23	120	29.5	90.5	0	0
24	35.5	35.5	0	0	0

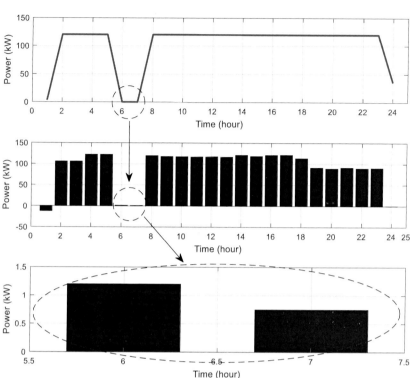

FIGURE 3.34 Grid and battery swapping station powers after an outage of the line at hours 6–7.

diesel generator and solar system. The swapping station does not inject power into the system because the produced power by other resources is enough to handle the loads. At hour 6, the load and swapping station consume 5 kW and it is supplied by the diesel generator. At hour 7, the load and swapping station consume 5.75 kW, where, 5 kW is provided by the diesel generator and 0.75 kW is supplied by solar energy.

Fig. 3.34 shows the grid power and battery swapping station power following the outage of the line at hours 6 and 7. Although the battery swapping station does not produce power at hours 6 and 7, the consumed power by the station is properly regulated and reduced close to zero. Such charging scheduling assists the system to deal with outages and events.

3.5 Conclusions

In this chapter, the impacts of electric vehicles on energy management in homes and residential microgrids were discussed. The home integrated with the parking station, the residential microgrid with the charging station, and the residential microgrid integrated with the battery swapping station were modeled and discussed.

In the HEMS with parking station, it was assumed that the electric vehicles remain inside the parking station for several hours and during this time period, the HEMS is allowed to optimize their charging scheduling. The HEMS intended to minimize the daily energy cost by optimal operation of resources, optimal charging scheduling of electric vehicles as well as optimal adjustment of loads. In the HEMS with the parking station, the home was connected to the grid and the grid power took both positive and negative values. The load was modeled as fixed, interruptible, shiftable, and curtailable loads. The load curtailment was done to deal with the shortage of energy in the building. The electric vehicles in the parking station were allowed to be charged or discharged at various hours. When the electric vehicle was leaving the parking station it was fully charged. The daily energy cost was achieved as 55.965 $/day. The fixed load was supplied at all hours without any interruption, the interpretable load was operated for 6 consecutive hours at the initial hours of the day and the shiftable load was supplied at 4 various hours. At hours 13, 14, and 19, the home sent the surplus of energy to the grid for making revenue. At many hours like 4, 9, 13, and 19 the parking station operated as a resource and produced energy for the home. The biggest discharge process appeared at hour 19 when the home faced the high-priced electricity and HEMS was attempting to sell electricity to the external grid for making a profit. When the discharge capability of electric vehicles was not permitted, the daily energy cost increased by 10%. When the capacity of the line was limited to 15 kW, the home was not able to send energy to the grid and only received energy from the external grid. The diesel generator operation was increased to deal with the shortage of power caused by limited line capacity. The uncertainty in solar and load energy increased the energy cost by about 11% because the HEMS was forced to employ more resources in order to deal with the developed uncertainty. The electric vehicles were optimized to have a unique operating pattern under all scenarios of performance and their operation did not change from one scenario to another one.

In the residential microgrid with the electric vehicle charging station, the electric vehicles remained in the charging station for two hours (8 time-intervals) to get fully charged. The energy cost was optimized as 228.952 $/day. The electric vehicle charging station appeared as a resource and produced electricity at many time intervals. The maximum produced power by the electric vehicle charging station was 100 kW at time intervals 47 and 49. The power to the grid became negative in many time intervals which demonstrated selling power to the grid by the microgrid. The diesel generator was operated especially in on-peak loading time intervals. The electric vehicles were discharged and their batteries were depleted in some time intervals but they were afterward charged for having full energy when leaving the station. When the discharge capability was not allowed, the daily energy cost increased by about 5%. The system without discharge capability was not able to encounter the events. The operation of the microgrid following the outage of the line at time intervals 77−80 showed that the system without discharge capability cannot deal with such an event and has an infeasible operation. The stochastic model including scenarios of performance increased the daily energy cost by about 3.5%.

In the microgrid with the battery swapping station, the daily energy cost was achieved equal to 375.955 $/day. The adjustable loads such as interpretable and shiftable loads were scheduled to be operated at the hours with low-priced electricity. All the depleted batteries of entered electric vehicles were properly replaced with fully charged batteries while the rest of the batteries participated in the energy management system. When the discharge capability was not used, the daily energy cost increased by 5%. After the outage of the line at hours 19−20, the discharge capability helped the system to deal with the outage, whereas the system without discharge capability showed infeasible operation.

References

[1] Hosseini Salari A, Mirzaeinejad H, Fooladi Mahani M. A new control algorithm of regenerative braking management for energy efficiency and safety enhancement of electric vehicles. Energy Conversion and Management 2023;276:116564.
[2] Tian J, Liu X, Li S, Wei Z, Zhang X, Xiao G, et al. Lithium-ion battery health estimation with real-world data for electric vehicles. Energy 2023;270:126855.
[3] Mehrjerdi H, Hemmati R. Stochastic model for electric vehicle charging station integrated with wind energy. Sustainable Energy Technologies and Assessments 2020;37.
[4] https://www.comlight.com/post/what-is-level-1-2-3-ev-charging.
[5] Yang B, Li J, Shu H, Cai Z, Tang B, Huang X, et al. Recent advances of optimal sizing and location of charging stations: a critical overview. International Journal of Energy Research 2022;46(13):17899−925.
[6] Mehrjerdi H, Hemmati R. Electric vehicle charging station with multilevel charging infrastructure and hybrid solar-battery-diesel generation incorporating comfort of drivers. Journal of Energy Storage 2019;26.
[7] Mehrjerdi H. Off-grid solar powered charging station for electric and hydrogen vehicles including fuel cell and hydrogen storage. International Journal of Hydrogen Energy 2019;44(23):11574−83.
[8] Saboori H, Jadid S. Mobile battery-integrated charging station for reducing electric vehicles charging queue and cost via renewable energy curtailment recovery. International Journal of Energy Research 2022;46(2):1077−93.

[9] Zhan W, Wang Z, Zhang L, Liu P, Cui D, Dorrell DG. A review of siting, sizing, optimal scheduling, and cost- benefit analysis for battery swapping stations. Energy. 2022;258:124723.

[10] Mehrjerdi H, Hemmati R. Coordination of vehicle-to-home and renewable capacity resources for energy management in resilience and self-healing building. Renewable Energy 2020;146:568−79.

[11] Hemmati R, Mehrjerdi H. Investment deferral by optimal utilizing vehicle to grid in solar powered active distribution networks. Journal of Energy Storage 2020;30.

[12] Mehrjerdi H, Hemmati R, Mahdavi S, Shafie-khah M, Catalao JP. Multi-carrier microgrid operation model using stochastic mixed integer linear programming. IEEE Transactions on Industrial Informatics 2022;18(7):4674−87.

[13] Panah PG, Bornapour M, Hemmati R, Guerrero JM. Charging station stochastic programming for hydrogen/battery electric buses using multi-criteria crow search algorithm. Renewable & Sustainable Energy Reviews 2021;144.

[14] Amiri SS, Jadid S, Saboori H. Multi-objective optimum charging management of electric vehicles through battery swapping stations. Energy. 2018;165:549−62.

[15] Hemmati R, Mahdavi S. Hybrid renewable/nonrenewable/storage resources in electrical grid considering active- reactive losses and depth of discharge. International Journal of Energy Research 2021;45(14):20384−99.

Capacity expansion planning in microgrids

DOI: https://doi.org/10.1016/B978-0-443-23728-7.00003-5
141
© 2024 Elsevier Inc. All rights reserved.

Nomenclature

Sets and indexes
t Index of daily time intervals
T Set of daily time intervals
y Index of years in planning horizon
Y Set of years in planning horizon
s Index of scenarios of performance
S Set of scenarios of performance

Parameters and variables
ao_{cb} Annual operational-maintenance cost of each battery capacity (\$/year)
ao_{pb} Annual operational-maintenance cost of each battery power (\$/year)
ao_{line} Annual operational-maintenance cost of each line (\$/year)
ao_{solar} Annual operational-maintenance cost of each solar system (\$/year)
ao_{wind} Annual operational-maintenance cost of each wind system (\$/year)
$C_{ele}^{t,y}$ Electricity price in deterministic model (\$/kWh)
$C_{ele}^{t,y,s}$ Electricity price under each scenario (\$/kWh)
$C_{fuel}^{t,y,s}$ Fuel price under each scenario (\$/kWh)
$C_{fuel}^{t,y}$ Fuel price in deterministic model (\$/kWh)
dr Discount rate (%)
dt Duration of time interval (hour)
E_b^0 Initial energy inside the battery (MWh)
E_{rb}^0 Capacity of battery in year zero (MWh)
EAC Equivalent annual cost (\$/year)
E_{rb}^y Rated capacity of battery (MWh)
$E_b^{t,y}$ Stored energy by battery (MWh)
F_{dy} Factor to convert daily cost to annual cost (equal to 365)
F_{mk} Factor to convert MW to kW (equal to 1000)
I_{cp} Investment cost of each battery capacity (\$)
I_{diesel} Investment cost of each diesel generator (\$)
I_{line} Investment cost of each line (\$)
I_{pb} Investment cost of each battery power (\$)
I_{solar} Investment cost of each solar system (\$)
I_{wind} Investment cost of each wind system (\$)
lt Asset lifetime (year)
NPV Net present value
O_{pr}^s Probability of occurrence for each scenario of performance
P_{pb}^0 Power of battery in year zero (MW)
P_{diesel}^0 Power of diesel generator in year zero (MW)
P_{line}^0 Power of line in year zero (MW)
P_{solar}^0 Power of solar system in year zero (MW)
P_{wind}^0 Power of wind system in year zero (MW)
P_{rb}^y Rated power of battery (MW)
$P_{rdiesel}^y$ Rated power of diesel generator (MW)
P_{rload}^y Rated power of load (MW)

P_{rline}^{y}	Rated power of line (MW)
P_{rsolar}^{y}	Rated power of solar system (MW)
P_{rwind}^{y}	Rated power of wind system (MW)
$P_{cb}^{t,y}$	Charged power to battery (MW)
$P_{db}^{t,y}$	Discharged power from battery (MW)
$P_{diesel}^{t,y}$	Produced power by diesel generator (MW)
$P_{grid}^{t,y}$	Power from grid in deterministic model (MW)
$P_{grid}^{t,y,s}$	Power from grid under each scenario (MW)
P_{load}^{t}	Consumed power by load in deterministic model (MW)
$P_{load}^{t,s}$	Consumed power by load under each scenario (MW)
P_{solar}^{t}	Produced power by solar system in deterministic model (MW)
$P_{solar}^{t,s}$	Produced power by solar system under each scenario (MW)
P_{wind}^{t}	Produced power by wind system in deterministic model (MW)
$P_{wind}^{t,s}$	Produced power by wind system under each scenario (MW)
TIC	Total investment cost ($)
ua_{diesel}^{y}	Vector showing installed diesel generators in each year
ua_{line}^{y}	Vector showing installed lines in each year
ua_{pb}^{y}	Vector showing installed battery powers in each year
ua_{cp}^{y}	Vector showing installed battery capacities in each year
ua_{solar}^{y}	Vector showing installed solar systems in each year
ua_{wind}^{y}	Vector showing installed wind systems in each year
ut_{pb}^{y}	Vector showing cumulative installed battery powers until each year
ut_{cb}^{y}	Vector showing cumulative installed battery capacities until each year
ut_{line}^{y}	Vector showing cumulative installed lines until each year
ut_{solar}^{y}	Vector showing cumulative installed solar systems until each year
ut_{wind}^{y}	Vector showing cumulative installed wind systems until each year
$u_{d}^{t,y}$	Binary variable showing battery discharging state [0,1]
$u_{c}^{t,y}$	Binary variable showing battery charging state [0,1]
Z_{of}	Total cost of planning ($/year)
Za_{ict}	Annual investment cost of planning ($/year)
Za_{oct}	Annual operational cost of planning ($/year)
Zi_{cp}	Annual investment cost on all battery capacities ($/year)
Zi_{diesel}	Annual investment cost on all diesel generators ($/year)
Zi_{line}	Annual investment cost on all lines ($/year)
Zi_{pb}	Annual investment cost on all battery powers ($/year)
Zi_{solar}	Annual investment cost on all solar systems ($/year)
Zi_{wind}	Annual investment cost on all wind systems ($/year)
Zo_{cb}	Annual operational-maintenance cost of all battery capacities ($/year)
Zo_{diesel}	Annual operational-maintenance cost of all diesel generators ($/year)
Zo_{grid}	Annual operational cost of purchasing electricity from grid ($/year)
Zo_{line}	Annual operational-maintenance cost of all lines ($/year)
Zo_{pb}	Annual operational-maintenance cost of all battery powers ($/year)
Zo_{solar}	Annual operational-maintenance cost of all solar systems ($/year)
Zo_{wind}	Annual operational-maintenance cost of all wind systems ($/year)
η_{b}	Efficiency of battery (%)

4.1 Introduction

In the electrical networks, expansion planning is carried out to deal with load growth in the future. The expansion planning may be done in the generation, transmission, or distribution sectors in the electrical networks. In the generation expansion planning, the new electric power plants are installed and added to the existing power stations to handle the load growth over the planning horizon which is typically 5–20 years. Various types of power stations like renewable units, fossil fuel units, hydro stations, and nuclear power plants may be involved in the planning. Such planning is carried out in both deregulated networks and traditional power systems. In the deregulated networks, the planning is conducted to maximize the benefit of private participants and stakeholders but in traditional networks, the generation expansion planning usually attempts to minimize the investment and operational costs of generating systems. The technical constraints of electricity generation, reliability, resilience, reserve margin, ramp rate, the uncertainty of renewables, uncertainty in prices, and environmental pollution are some key subjects that are taken into account in generation expansion planning [1]. The generation expansion planning finds the location, capacity, time of installation, technology, and number of newly installed power stations in the network [2].

Transmission expansion planning is another part of network expansion planning in which the limitations of transmission lines are resolved by installing new lines or reinforcing the available lines. In transmission expansion planning, the technical constraints of the grid like feasibility of power flow, voltage limits, reliability, line capacity limitations, and losses are usually taken into account [3]. This planning the same as generation expansion planning can be carried out in both the deregulated and traditional networks. The transmission expansion planning determines the location, capacity, time of installation, type, and number of newly installed lines in the network [4].

The distribution network expansion planning is a planning similar to the transmission expansion planning but it is carried out on the distribution grids. The distribution network expansion planning denotes the location, capacity, time of installation, type, and number of newly installed lines in each corridor of the distribution grid [5]. Since the radiality of the distribution grid is necessary, one of the constraints in the distribution network expansion planning is to keep the grid radiality following installing new lines. The technical constraints of the distribution grid-like feasibility of power flow, voltage range, line capacity, and reliability are considered by the planning [6].

Coordinated planning may also be presented for the expansion of generation, transmission, and distribution sections together [7]. In the coordinated transmission and generation expansion planning, both the generating units and transmission lines are expanded to handle future load growth [8]. The collaborative expansion planning for transmission and distribution networks is also addressed to resolve the issues of both grids at the same time [9]. Novel technologies like energy storage systems [10], electric vehicle charging stations [11], and flexible AC transmission system (FACTS) devices are widely used in the expansion planning of power systems [12].

The microgrid is a structure similar to the electrical networks but smaller in size and capacity. The microgrid is often integrated with renewable resources like solar generating systems and wind turbines. The fossil-fuel generating systems like diesel generators and micro gas turbines may also be installed in the microgrid for emergency operation. Energy storage systems are an inevitable part of the microgrid. The microgrid is connected to the external distribution grid by a line [13]. The expansion planning in the microgrid can be carried out similarly to the other sectors of electrical networks. In the microgrid expansion planning, all the resources as well as the connecting line to the external grid may be subject to expansion. Since there are many resources in the microgrid, long-term expansion planning may be carried out together with short-term operating scheduling. Such coordinated long-term planning and short-term scheduling provide better results compared to the classic models that only deal with long-term planning. Since the microgrids are directly connected to the distribution grid, they can be combined with the distribution network expansion planning to make positive impacts on the distribution grid [14]. Subject matters such as environmental pollution, reliability, uncertainty in parameters, demand response programs, thermal loads, multicarrier energy systems, and electric vehicle charging stations could be taken into account in the microgrid expansion planning [15].

In this chapter, the expansion planning is carried out on a residential microgrid. The resources in the microgrid are the wind generating system, solar panels, and diesel generator. The battery energy storage system is also integrated into the microgrid. In the energy storage system, the capacity is denoted by the rated capacity of series-parallel batteries, and the power is determined by the rated power of DC−AC converters between the batteries and the AC grid. The microgrid is connected to the upstream distribution network by a line. A 6-year planning horizon is taken into account and a typical load growth is defined over the planning horizon. The microgrid needs to be expanded to deal with such load growth. Both the resources and line are subject to expansion. The objective function of expansion planning is to minimize the annualized planning cost including the annual operational cost plus the annual investment cost. All the costs are annualized for having a similar unit in $/year as well as they are discounted back to the present year as net present value. While the objective function of long-term planning is to minimize the planning cost over the planning horizon, the short-term operating scheduling is optimized for all resources in all years through the planning horizon. In other words, the optimal 24-hour operating pattern is achieved for all resources in all years. Such short-term operating scheduling makes positive impacts on long-term expansion planning and decreases the planning cost because the load power at many hours especially at on-peak hours is supplied by local resources. Such a matched long-term planning and short-term scheduling is modeled as a mixed integer linear programming and solved by General Algebraic Modeling System (GAMS) software. An investigation is carried out on the impacts of prices in the planning and the expansion problem is simulated taking into account various prices for battery and line. The battery lifespan and operational cost are also changed in order to evaluate their effects on the plan. The resilient planning under the outage of elements is investigated as well. In such a paradigm, the key elements such as the line between the microgrid and the external network are exposed to the outage and the expansion planning is conducted taking into account such outages. During outage of line, the microgrid operates as islanded and it must supply all the loads with its own resources. Such points change the expansion planning entirely and resilient expansion planning completely differs from classic planning. Moreover, stochastic expansion planning is addressed by considering wind energy, solar energy, load power, electricity price, and fuel price as uncertain parameters and modeling them by a set of scenarios of performance.

4.1.1 Some research gaps in long-term expansion planning

The main research gaps in long-term expansion planning may be summarized as follows:

Future of electricity generation: Expansion planning is a long-term problem that deals with issues of the networks in the next 10 or 20 years. As a result, it must consider the future of electricity generation. In the next 20 years, fossil fuel energy resources may face barriers and new energies would be replaced. The sustainability of energy resources is also a matter of great concern. The new and sustainable energy resources need to be investigated and included in the long-term expansion planning. Hydrogen, biofuels, biogases, renewables, and hydro energy are some examples.

Resilience and self-healing grids: In the future years, the electrical grids will be broadly faced with physical attacks, cyber-attacks, and natural disasters such as flooding, drought, earthquake, hurricane, tornados, wildfires, and tsunamis. The resilient and self-heading networks are essential for reliable and stable operation amid those challenges. The leading solutions are required to design self-healing grids to enhance reliability today and to meet future necessities as well.

Future technologies: In the future years, the current technologies will get mature and come to widespread use. The technologies such as mobile energy storage systems, large-scale electric vehicle charging stations, battery swapping stations, seasonally transferred energy resources, mobile electric vehicle charging stations, smart cities, and smart energy systems will become mature technologies. They could be included in long-term planning.

4.2 Microgrid expansion planning

In microgrid expansion planning, the available (existing) and candidate resources are included in the model to find optimal results over the planning horizon. As shown in Fig. 4.1, the microgrid with available resources is connected to the grid through a line. The available resources and the line are candidate elements for expansion by microgrid expansion planning. The planning installs new resources in the microgrid by finding the optimal capacity, number, and installation time for new resources as follows [14]:

- Finding optimal power, number, and installation time for new wind turbines.
- Finding optimal power, number, and installation time for new solar panels.
- Finding optimal power, number, and installation time for new diesel generators.
- Finding optimal capacity, number, and installation time for new lines.
- Finding optimal capacity, number, and installation time for new batteries in the energy storage system.
- Finding optimal power, number, and installation time for new DC−AC converters in the energy storage system.

These new resources are installed in different years over the planning horizon and added to the existing resources and lines. The objective of the planning is to minimize the operational and investment costs as follows [14];

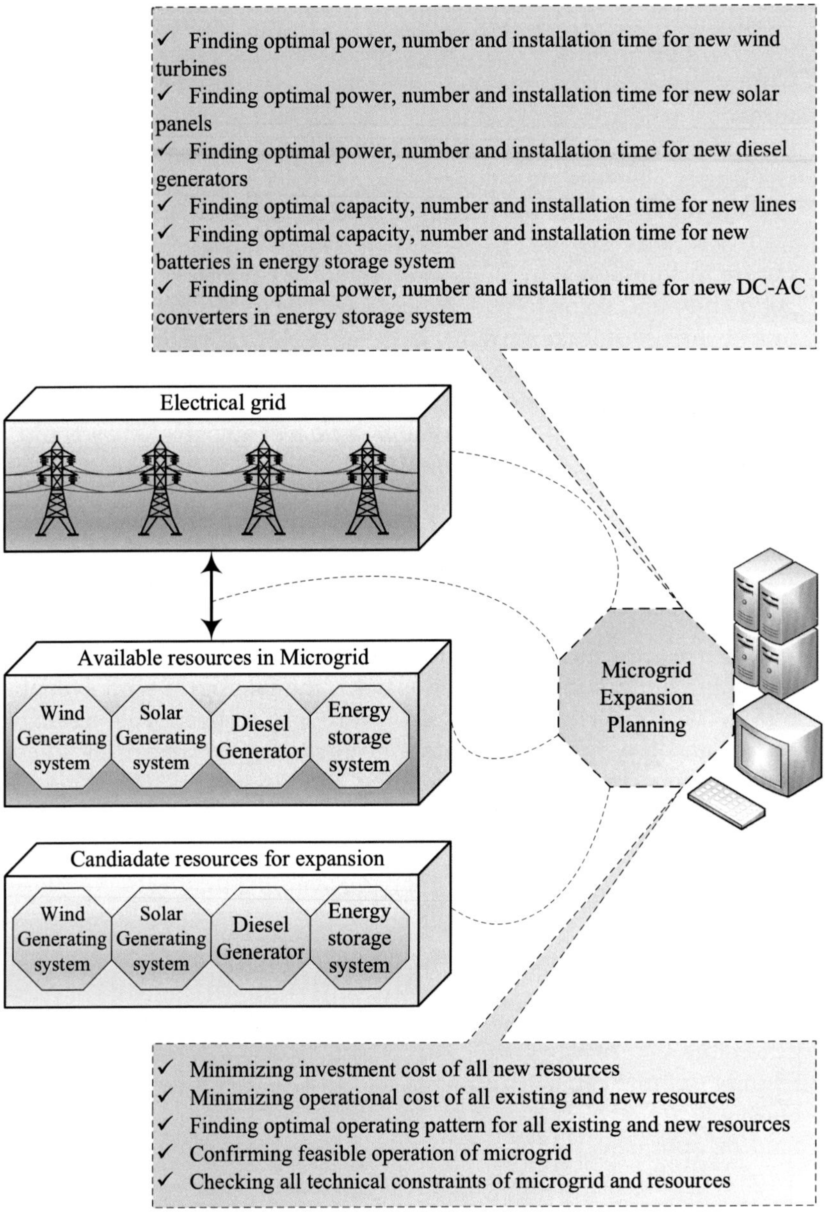

✓ Finding optimal power, number and installation time for new wind turbines
✓ Finding optimal power, number and installation time for new solar panels
✓ Finding optimal power, number and installation time for new diesel generators
✓ Finding optimal capacity, number and installation time for new lines
✓ Finding optimal capacity, number and installation time for new batteries in energy storage system
✓ Finding optimal power, number and installation time for new DC-AC converters in energy storage system

Electrical grid

Available resources in Microgrid

| Wind Generating system | Solar Generating system | Diesel Generator | Energy storage system |

Microgrid Expansion Planning

Candiadate resources for expansion

| Wind Generating system | Solar Generating system | Diesel Generator | Energy storage system |

✓ Minimizing investment cost of all new resources
✓ Minimizing operational cost of all existing and new resources
✓ Finding optimal operating pattern for all existing and new resources
✓ Confirming feasible operation of microgrid
✓ Checking all technical constraints of microgrid and resources

FIGURE 4.1　Long-term expansion planning problem.

- Minimizing investment cost of all new resources.
- Minimizing the operational-maintenance cost of all existing and new resources.

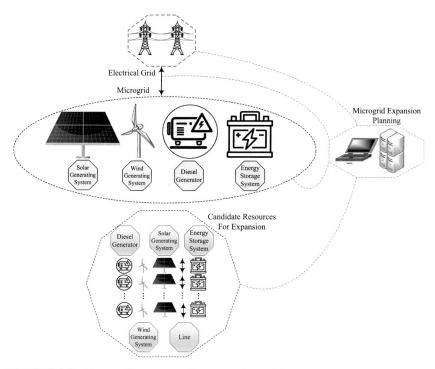

FIGURE 4.2 Microgrid expansion planning with candidate resources.

The operational costs are usually achieved in $/day but the investment costs are attained per asset lifetime. They do not have the same unit and need to be presented as equivalent annual cost in $/year. The planning also finds the optimal short-term operating scheduling for resources as follows:

- Finding optimal operating patterns for all existing and new resources.
- Confirming feasible operation of the microgrid at each hour.
- Ensuring all technical constraints of the microgrid and resources.

Fig. 4.2 shows the proposed microgrid with a wind generating system, solar panels, diesel generator, and battery energy storage system as local resources. The microgrid is connected to the external distribution grid through a line. The candidate resources are selected by planning and installed on the microgrid. The line between the microgrid and the external grid is also considered a candidate for expansion.

In the energy storage system, two parameters of power and capacity are defined for characterizing the energy storage system operation. As designated in Fig. 4.3, the capacity of the energy storage system is specified by the capacity of all batteries which are used in the energy storage

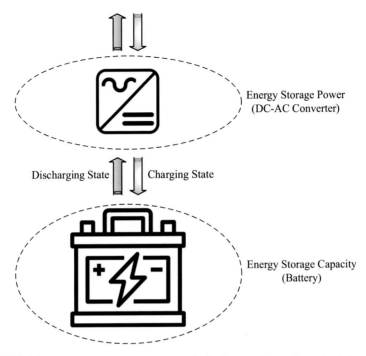

Energy Storage Power
(DC-AC Converter)

Discharging State   Charging State

Energy Storage Capacity
(Battery)

FIGURE 4.3   Battery energy storage system including rated capacity and rated power.

system. The capacity is measured per kWh. In the actual energy storage systems, the batteries are connected in seriesparallel to achieve the desired capacity, voltage, and current. The power of the energy storage system is determined by the rated power of DC−AC converters between the batteries and the AC grid. The rated power can be increased by installing larger converters or reinforcing the available converters by means of installing new parallel converters. Both the charging and discharging powers passing through the converter should be less than or equal to the rated power of the converter. The stored energy by the energy storage system should be less than or equal to the rated capacity of all batteries which are equipped in the energy storage system [16].

4.3 Expansion planning formulation

In long-term expansion planning, there are two types of costs including operational-maintenance costs and investment costs. The operational costs are obtained per day as $/day and they show the daily fuel cost or daily electricity costs. On the other hand, the investment costs are obtained per asset lifetime. These two types of costs cannot be added because they do not have the same unit. In order to sum up these costs, both of them are converted to the annual cost and presented as $/year. The daily costs are elaborated over the year to calculate the annual operational cost in "$/year." They are multiplied by the number of days (i.e., 365) to be presented as annual. The investment costs are converted to the equivalent annual cost (EAC) in "$/year" by Eq. (4.1). In this relationship, the total investment cost in "$/lifetime" is converted to the equivalent annual cost in "$/year." As well, all the costs in every year are discounted back to the present year as net present value (NPV) which is expressed by Eq. (4.2) [14].

$$EAC = \left[\frac{dr \times (1+dr)^{lt}}{(1+dr)^{lt} - 1} \right] \times TIC \tag{4.1}$$

$$NPV = \frac{1}{(1+dr)^y} \tag{4.2}$$

The purchased power from the upstream grid in all years of the planning horizon is calculated by Eq. (4.3). The operation of the diesel generator needs fuel and the related operational cost in all years of the planning horizon is expressed by Eq. (4.4). The operational and maintenance costs of solar and wind systems in all years of the planning horizon are given by Eqs. (4.5) and (4.6), receptively. The maintenance cost of the line between the microgrid and the external grid is denoted as Eq. (4.7). The operational and maintenance costs of battery power (i.e., converter between the storage device and the grid) and the battery capacity (storage device) in all years of planning horizon are given by Eqs. (4.8) and (4.9), receptively. It is seen that all the costs are discounted back to the present year as net present value [14].

$$Zo_{grid} = \sum_{y=1}^{Y} \left(\sum_{t=1}^{T} \left(P_{grid}^{t,y} \times F_{mk} \times C_{ele}^{t,y} \right) \times F_{dy} \times NPV \right) \tag{4.3}$$

$$Zo_{diesel} = \sum_{y=1}^{Y} \left(\sum_{t=1}^{T} \left(P_{diesel}^{t,y} \times F_{mk} \times C_{fuel}^{t,y} \right) \times F_{dy} \times NPV \right) \tag{4.4}$$

$$Zo_{solar} = \sum_{y=1}^{Y} \left(ut_{solar}^{y} \times ao_{solar} \times NPV \right) \tag{4.5}$$

$$Zo_{wind} = \sum_{y=1}^{Y} \left(ut_{wind}^{y} \times ao_{wind} \times NPV \right) \tag{4.6}$$

$$Zo_{line} = \sum_{y=1}^{Y} \left(ut_{line}^{y} \times ao_{line} \times NPV \right) \tag{4.7}$$

$$Zo_{pb} = \sum_{y=1}^{Y} \left(ut_{pb}^{y} \times ao_{pb} \times NPV \right) \tag{4.8}$$

$$Zo_{cb} = \sum_{y=1}^{Y} \left(ut_{cb}^{y} \times ao_{cb} \times NPV \right) \tag{4.9}$$

The annual investment cost of the solar system is presented by Eq. (4.10). It is seen that the investment costs are converted to the equivalent annual cost. As well, all the costs are discounted back to the present year as net present value. The annual investment cost of the wind system is presented in Eq. (4.11). The cost of installing new lines is presented in Eq. (4.12). The investment cost of battery power and battery capacity are given by Eqs. (4.13) and (4.14), receptively. The annual investment cost of diesel generators is presented by Eq. (4.15).

$$Zi_{solar} = \sum_{y=1}^{Y} \left(ua_{solar}^{y} \times I_{solar} \times NPV \right) \times EAC \tag{4.10}$$

$$Zi_{wind} = \sum_{y=1}^{Y} \left(ua_{wind}^{y} \times I_{wind} \times NPV \right) \times EAC \tag{4.11}$$

$$Zi_{line} = \sum_{y=1}^{Y} \left(ua_{line}^{y} \times I_{line} \times NPV \right) \times EAC \tag{4.12}$$

$$Zi_{pb} = \sum_{y=1}^{Y} \left(ua_{pb}^{y} \times I_{pb} \times NPV \right) \times EAC \tag{4.13}$$

$$Zi_{cp} = \sum_{y=1}^{Y} \left(ua_{cp}^{y} \times I_{cp} \times NPV \right) \times EAC \tag{4.14}$$

$$Zi_{diesel} = \sum_{y=1}^{Y} \left(ua_{diesel}^{y} \times I_{diesel} \times NPV \right) \times EAC \tag{4.15}$$

The annual operational cost of the microgrid is calculated by Eq. (4.16) and the annual investment cost of the microgrid is expressed by Eq. (4.17). The total cost including operational and investment costs is given by Eq. (4.18). This final cost is considered the objective function of planning and is minimized. It is the annualized cost in $/year.

$$Za_{oct} = \left(Zo_{grid} + Zo_{diesel} + Zo_{solar} + Zo_{wind} + Zo_{line} + Zo_{pb} + Zo_{cb} \right) \quad (4.16)$$

$$Za_{ict} = \left(Zi_{solar} + Zi_{wind} + Zi_{line} + Zi_{pb} + Zi_{cp} + Zi_{diesel} \right) \quad (4.17)$$

$$Z_{of} = Za_{oct} + Za_{ict} \quad (4.18)$$

The exchanged power with the external grid is formulated by Eq. (4.19). The first part shows the consumed powers and the second part indicates the generated powers. If energy generation is more than energy consumption, the exchanged power becomes negative which means selling power to the grid by the microgrid. The microgrid can trade energy to the grid for making revenue. Generally, when electricity in the grid is expensive, it is not economical to buy such pricy electricity from the grid. In such occasions, the microgrid should attempt to supply the loads with its own resources. In such times, the microgrid can also sell the surplus of energy to the grid for making higher revenue. The load and charging power of the battery are assumed as energy consumers. The solar system, wind unit, diesel generator, and discharging power of the battery are regarded as energy producers. The power balance must be approved at all hours of all years over the planning horizon.

$$P_{grid}^{t,y} = \left[P_{load}^{t} \times P_{rload}^{y} + P_{cb}^{t,y} \right] - \left[P_{db}^{t,y} + P_{solar}^{t} \times P_{rsolar}^{y} + P_{wind}^{t} \times P_{rwind}^{y} + P_{diesel}^{t,y} \right]$$
$$\forall t \in T$$
$$\forall y \in Y$$

$$(4.19)$$

The installed solar systems in each year are optimized by the planning. In each year, the solar systems in the microgrid are calculated by Eq. (4.20) which shows the cumulative number of solar systems. In other words, the available solar systems in each year are equal to the installed solar systems in the previous years plus the new installed solar systems in the present year. The rated power of the solar system in each year is calculated by Eq. (4.21). It specifies the cumulative number of solar systems in each year added to the existing solar system in year zero before starting the planning. Similar to the solar system, the cumulative number of wind systems and the rated power of wind turbines in each year are calculated by Eqs. (4.22) and (4.23).

$$\begin{cases} ut_{solar}^{y} = ua_{solar}^{y} & \forall y = 1 \\ ut_{solar}^{y} = ut_{solar}^{y-1} + ua_{solar}^{y} & \forall y \in Y, y \notin 1 \end{cases} \quad (4.20)$$

$$P^y_{rsolar} = P^0_{solar} \times \left(1 + ut^y_{solar}\right)$$
$$\forall y \in Y \tag{4.21}$$

$$\begin{cases} ut^y_{wind} = ua^y_{wind} & \forall y = 1 \\ ut^y_{wind} = ut^{y-1}_{wind} + ua^y_{wind} & \forall y \in Y, y \notin 1 \end{cases} \tag{4.22}$$

$$P^y_{rwind} = P^0_{wind} \times \left(1 + ut^y_{wind}\right)$$
$$\forall y \in Y \tag{4.23}$$

The installed lines in each year, the cumulative number of lines in each year, and the rated capacity of lines in each year are calculated through Eqs. (4.24) and (4.25). The exchanged power with the grid must be less than the rated capacity of lines as presented by Eq. (4.26).

$$\begin{cases} ut^y_{line} = ua^y_{line} & \forall y = 1 \\ ut^y_{line} = ut^{y-1}_{line} + ua^y_{line} & \forall y \in Y, y \notin 1 \end{cases} \tag{4.24}$$

$$P^y_{rline} = P^0_{line} \times \left(1 + ut^y_{line}\right)$$
$$\forall y \in Y \tag{4.25}$$

$$\begin{cases} P^{t,y}_{grid} \leq + P^y_{rline} \\ P^{t,y}_{grid} \geq - P^y_{rline} \end{cases}$$
$$\forall y \in Y$$
$$\forall t \in T \tag{4.26}$$

The installed diesel generators in each year, the cumulative number of diesel generators in each year, and the rated capacity of diesel generators in each year are denoted by Eqs. (4.27) and (4.28). The produced power by diesel generators at all hours of all years must be less than the rated power of diesel generators as presented in Eq. (4.29).

$$\begin{cases} ut^y_{diesel} = ua^y_{diesel} & \forall y = 1 \\ ut^y_{diesel} = ut^{y-1}_{diesel} + ua^y_{diesel} & \forall y \in Y, y \notin 1 \end{cases} \tag{4.27}$$

$$P^y_{rdiesel} = P^0_{diesel} \times \left(1 + ut^y_{diesel}\right)$$
$$\forall y \in Y \tag{4.28}$$

$$P^{t,y}_{diesel} \leq P^y_{rdiesel}$$
$$\forall t \in T$$
$$\forall y \in Y \tag{4.29}$$

The battery energy storage system is used to shift energy by charging energy when electricity is not needed and discharging energy when it is required. At each hour, the battery operates in either the charring or discharging state. This point is modeled by binary variables in Eq. (4.30). As constrained by Eq. (4.30), only one of the binary variables can be one and the other one is forced to be zero. Accordingly, at each hour only one of the charring or discharging powers in Eqs. (4.31) and (4.32) can get value and the other one is zero. The charging or discharging powers should be less than the rated power of the converter between the battery and grid as modeled by Eqs. (4.33) and (4.34).

$$u_c^{t,y} + u_d^{t,y} \le 1$$
$$\forall t \in T$$
$$\forall y \in Y \tag{4.30}$$

$$P_{cb}^{t,y} \le u_c^{t,y} \times Bm$$
$$\forall t \in T$$
$$\forall y \in Y \tag{4.31}$$

$$P_{db}^{t,y} \le u_d^{t,y} \times Bm$$
$$\forall t \in T$$
$$\forall y \in Y \tag{4.32}$$

$$P_{cb}^{t,y} \le P_{rb}^{y}$$
$$\forall t \in T$$
$$\forall y \in Y \tag{4.33}$$

$$P_{db}^{t,y} \le P_{rb}^{y}$$
$$\forall t \in T$$
$$\forall y \in Y \tag{4.34}$$

The converter between the battery and grid can be reinforced for increasing capacity by installing new parallel converters. The installed converters (i.e., battery power) in each year, the cumulative number of converters in each year, and the rated power of converters in each year are denoted by Eqs. (4.35) and (4.36).

$$\begin{cases} ut_{pb}^{y} = ua_{pb}^{y} & \forall y = 1 \\ ut_{pb}^{y} = ut_{pb}^{y-1} + ua_{pb}^{y} & \forall y \in Y, y \notin 1 \end{cases} \tag{4.35}$$

$$P_{rpb}^{y} = P_{pb}^{0} \times \left(1 + ut_{pb}^{y}\right)$$
$$\forall y \in Y \tag{4.36}$$

The stored energy by the battery at each hour is calculated as Eq. (4.37) and the initial energy inside the battery at hour zero (one hour before starting planning) is denoted by Eq. (4.38). The stored energy inside the battery should be less than the rated capacity of the battery as modeled by Eq. (4.39). The rated capacity of the battery in each year is obtained by Eqs. (4.40) and (4.41), where, the installed batteries in each year, the cumulative number of batteries in each year, and the rated capacity of batteries in each year are denoted.

$$E_b^{t,y} = E_b^{t,y} + \left[P_{cb}^{t,y} - \frac{P_{db}^{t,y}}{\eta_b} \right] \times dt$$

$$\forall t \in T, t \notin 1$$
$$\forall y \in Y$$

(4.37)

$$E_b^{t,y} = E_b^0 + \left[P_{cb}^{t,y} - \frac{P_{db}^{t,y}}{\eta_b} \right] \times dt$$

$$\forall t \in [1]$$
$$\forall y \in Y$$

(4.38)

$$E_b^{t,y} \le E_{rb}^y$$
$$\forall t \in T$$
$$\forall y \in Y$$

(4.39)

$$\begin{cases} ut_{cb}^y = ua_{cb}^y & \forall y = 1 \\ ut_{cb}^y = ut_{cb}^{y-1} + ua_{cb}^y & \forall y \in Y, y \notin 1 \end{cases}$$

(4.40)

$$E_{rb}^y = E_{rb}^0 \times \left(1 + ut_{cb}^y\right)$$
$$\forall y \in Y$$

(4.41)

4.4 Test system data

The introduced test system is equipped with battery, load, wind, and solar systems and it is connected to the external grid through a line. The data of resources are listed in Table 4.1 [14]. These resources exist in the microgrid and the purpose is to install new resources to deal with load growth during the planning horizon which is taken equal to 6 years. Each day is modeled by 24 hours. The fuel price for the diesel generator is 0.1 \$/kWh. Table 4.2 shows the load and price growth over the planning horizon [14]. It is clear that the microgrid faces enormous load growth and it needs to expand the existing resources by installing new resources in order to deal with such load evolution. Table 4.3 lists the daily patterns for load, renewable energy, and electricity price. The initial energy of the battery is equal to zero and the efficiency of the

TABLE 4.1　Available resources before starting expansion planning.

	Rated power-capacity	Investment cost ($)	Lifetime (year)
Solar system	0.2 MW	250,000	10
Wind system	0.2 MW	350,000	8
Line to grid	0.8 MW	100,000	10
Diesel generator	0.4 MW	30,000	5
Power of battery	0.1 MW	30,000	4
Capacity of battery	0.4 MWh	30,000	4

TABLE 4.2　Load and price growth over the planning horizon.

Year No.	Price (%)	Load (%)
1	100	100
2	105	120
3	110	135
4	115	155
5	120	170
6	125	195

battery is considered 100%. The annual maintenance-operational cost of resources (in $/year) is considered as 5% of the annual investment cost in each year [14].

4.5 Expansion planning on the test system

The developed model for microgrid expansion planning is simulated on the introduced test case and the results are listed in Table 4.4. The resources in year zero demonstrate the existing resources in the microgrid before starting the expansion planning. The installed resources in each year are shown and it is clear that the planning installs solar and wind systems more than the other resources. The lower investment-operational cost of solar—wind systems is the reason for such output. The planning does not install a new diesel generator in the microgrid and a single new line is installed in year 4. Since the operational cost of the diesel generator is more than purchasing electricity from the grid,

TABLE 4.3   Daily patterns for load, renewable energy and electricity price.

Hour	Solar power (%)	Wind power (%)	Load power (%)	Electricity price ($/kWh)
1	0	10	50	0.05
2	0	10	50	0.05
3	0	10	50	0.05
4	0	10	50	0.05
5	0	10	50	0.05
6	0	10	50	0.05
7	0	30	50	0.05
8	0	30	50	0.05
9	50	50	60	0.06
10	50	50	60	0.06
11	50	50	60	0.06
12	50	50	60	0.06
13	100	100	70	0.07
14	100	100	70	0.07
15	100	100	70	0.07
16	100	100	70	0.07
17	60	70	100	0.10
18	60	70	100	0.10
19	60	70	100	0.10
20	60	70	100	0.10
21	0	10	90	0.09
22	0	10	90	0.09
23	0	10	90	0.09
24	0	10	90	0.09

the planning prefers to buy electricity from the grid rather than instal-ling new diesel generators and utilizing them. It should be noted that there is one diesel generator in the microgrid which is related to year zero. The power of the battery is not expanded too much and only two new converters are equipped to increase the battery power, but the capacity of the battery is expanded by installing four new batteries over

TABLE 4.4 Number of installed resources in each year.

Resource	Available resources in year 0	New installed resources						
		Year 1	Year 2	Year 3	Year 4	Year 5	Year 6
Solar system	1	1	1	1	1	1	1
Wind system	1	1	1	1	1	1	1
Line	1	0	1	0	0	0	0
Diesel generator	1	0	0	0	0	0	0
Power of battery	1	1	1	0	0	0	0
Capacity of battery	1	1	1	1	1	1	0

TABLE 4.5 Total number of available resources in each year.

Resource	Year 0	Year 1	Year 2	Year 3	Year 4	Year 5	Year 6
Solar system	1	2	3	4	5	6	7
Wind system	1	2	3	4	5	6	7
Line	1	1	2	2	2	2	2
Diesel generator	1	1	1	1	1	1	1
Power of battery	1	2	3	3	3	3	3
Capacity of battery	1	2	3	4	5	6	6

the planning horizon. Such results show that the microgrid needs a large capacity for storing energy.

Table 4.5 presents the total number of available resources in each year which is obtained by adding the newly installed resources in each year to the existing resources from previous years. The resources in year zero are also taken into account and must be added. In the last year of the planning, the microgrid is significantly equipped with various resources. The only diesel generator of the microgrid is related to year zero.

Table 4.6 presents the investment and operational costs which are annualized and presented in $/year. The total planning cost is equal to 1,566,832.887 $/year. The investment in the diesel generator is zero because the planning does not install new diesel generators in the

TABLE 4.6   Investment-operational costs and total planning cost.

Annual investment costs ($/year)	Investment cost of solar system ($/year)	164,322.55
	Investment cost of wind system ($/year)	274,818.362
	Investment cost of line system ($/year)	11,745.65
	Investment cost of battery power ($/year)	15,730.524
	Investment cost of battery capacity ($/year)	36,626.724
	Investment cost of diesel generator ($/year)	0
Annual operational costs ($/year)	Operational cost of solar system ($/year)	27,588.842
	Operational cost of wind system ($/year)	46,140.474
	Operational cost of line system ($/year)	2669.772
	Operational cost of battery power ($/year)	3891.092
	Operational cost of battery capacity ($/year)	6893.673
	Operational cost of diesel generator ($/year)	8343.024
	Cost of purchasing power from grid ($/year)	968,062.199
Total planning cost ($/year)		1,566,832.887

microgrid. The most investment is done in the wind turbines and the minimum investment is done in the battery power.

The diesel generator is operated only in year 1 at hours 21−24 as depicted in Fig. 4.4. The diesel generator is not operated too much and its operational cost is very low. It produces 60 kW at hours 21−24 during on-peak loading and it is switched off the rest of the time.

The equilibrium of power in the microgrid needs to be passed at all hours in all years of the planning horizon. The load and the battery charging power are the consumers. On the other hand, the battery discharging power, wind, solar, and diesel units are the power generators and the grid power works on both consuming and generating states. Tables 4.7−4.9 present the power balance in the microgrid in years 1, 4, and 6, respectively. It is clear that in all hours the produced power is equal to the consumed power. In year 1, the battery is charged when the wind and solar energy are in peak production or when the electricity is inexpensive. On the other hand, it is discharged when the electricity is

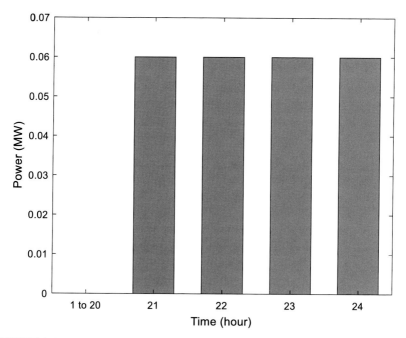

FIGURE 4.4   Diesel generator operation in year 1.

expensive such as hours 17–20. The diesel generator is also operated at hours with a shortage of energy production such as hours 21–24 when the solar energy is zero and wind energy is at the minimum level. A similar operation is seen in years 4 and 6. However, the level of power increases from year 1 to year 6, because the load is growing over the planning horizon and the new resources are installed. For instance, the peak load in year 1 is 1 MW while it is 1.95 MW in year 6. The peak power of wind and solar systems in year 1 is 0.4 MW while it is 1.4 MW in year 6. All the produced power by wind and solar systems at every hour is injected into the microgrid and the output power of those resources is not regulated or managed.

Table 4.10 presents the exchanged power with the grid at all hours in all years. In year 1, the microgrid sends power to the grid at hours 13–15. In the next years, new generating resources are installed and added to the microgrid, and its capability for producing power increases. As a result, it is able to send power to the grid in extra hours like hours 9–20. Such a point indicates that expansion planning not only reduces costs but also makes a profit by installing and operating resources optimally. The exchanged power with the grid increases from years 1 to 6 significantly due to loading-generating expansion in the microgrid over the planning horizon.

TABLE 4.7 Powers of resources in year 1 of the planning horizon.

Hour	Grid (MW)	Load (MW)	Battery charging (MW)	Battery discharging (MW)	Solar (MW)	Wind (MW)	Diesel generator (MW)
1	0.46	0.5	0	0	0	0.04	0
2	0.46	0.5	0	0	0	0.04	0
3	0.46	0.5	0	0	0	0.04	0
4	0.46	0.5	0	0	0	0.04	0
5	0.58	0.5	0.2	0	0	0.12	0
6	0.58	0.5	0.2	0	0	0.12	0
7	0.58	0.5	0.2	0	0	0.12	0
8	0.58	0.5	0.2	0	0	0.12	0
9	0.2	0.6	0	0	0.2	0.2	0
10	0	0.6	0	0.2	0.2	0.2	0
11	0.2	0.6	0	0	0.2	0.2	0
12	0.4	0.6	0.2	0	0.2	0.2	0
13	−0.3	0.7	0	0.2	0.4	0.4	0
14	−0.1	0.7	0	0	0.4	0.4	0
15	−0.1	0.7	0	0	0.4	0.4	0
16	0.1	0.7	0.2	0	0.4	0.4	0
17	0.28	1	0	0.2	0.24	0.28	0
18	0.28	1	0	0.2	0.24	0.28	0
19	0.28	1	0	0.2	0.24	0.28	0
20	0.28	1	0	0.2	0.24	0.28	0
21	0.8	0.9	0	0	0	0.04	0.06
22	0.8	0.9	0	0	0	0.04	0.06
23	0.8	0.9	0	0	0	0.04	0.06
24	0.8	0.9	0	0	0	0.04	0.06

Fig. 4.5 depicts the exchanged power with the grid in years 1 and 6 as two typical years. The negative parts of the power are clearly seen and such negative segments significantly increase from years 1 to 6. In year 6, the microgrid is equipped with many generating resources which enable the microgrid for producing and selling extra power to the grid at many hours, particularly during expensive time periods.

TABLE 4.8 Powers of resources in year 4 of the planning horizon.

Hour	Grid (MW)	Load (MW)	Battery charging (MW)	Battery discharging (MW)	Solar (MW)	Wind (MW)	Diesel generator (MW)
1	0.875	0.775	0.2	0	0	0.1	0
2	0.975	0.775	0.3	0	0	0.1	0
3	0.975	0.775	0.3	0	0	0.1	0
4	0.975	0.775	0.3	0	0	0.1	0
5	0.475	0.775	0	0	0	0.3	0
6	0.775	0.775	0.3	0	0	0.3	0
7	0.775	0.775	0.3	0	0	0.3	0
8	0.775	0.775	0.3	0	0	0.3	0
9	−0.07	0.93	0	0	0.5	0.5	0
10	−0.07	0.93	0	0	0.5	0.5	0
11	−0.07	0.93	0	0	0.5	0.5	0
12	−0.07	0.93	0	0	0.5	0.5	0
13	−0.785	1.085	0	0.3	1	1	0
14	−0.915	1.085	0	0	1	1	0
15	−0.915	1.085	0	0	1	1	0
16	−0.615	1.085	0.3	0	1	1	0
17	−0.05	1.55	0	0.3	0.6	0.7	0
18	−0.05	1.55	0	0.3	0.6	0.7	0
19	−0.05	1.55	0	0.3	0.6	0.7	0
20	−0.05	1.55	0	0.3	0.6	0.7	0
21	1.295	1.395	0	0	0	0.1	0
22	0.995	1.395	0	0.3	0	0.1	0
23	0.995	1.395	0	0.3	0	0.1	0
24	1.095	1.395	0	0.2	0	0.1	0

The line between the microgrid and the external grid is one of the resources which is expanded over the planning horizon. The flowed power through the line at hour 1 together with the line capacity in each year is presented in Fig. 4.6. It is clear that the flowed power increases from years 1 to 6 and the line needs to be expanded in year 2 because

TABLE 4.9 Powers of resources in year 6 of the planning horizon.

Hour	Grid (MW)	Load (MW)	Battery charging (MW)	Battery discharging (MW)	Solar (MW)	Wind (MW)	Diesel generator (MW)
1	1.135	0.975	0.3	0	0	0.14	0
2	1.135	0.975	0.3	0	0	0.14	0
3	1.135	0.975	0.3	0	0	0.14	0
4	1.135	0.975	0.3	0	0	0.14	0
5	0.855	0.975	0.3	0	0	0.42	0
6	0.855	0.975	0.3	0	0	0.42	0
7	0.855	0.975	0.3	0	0	0.42	0
8	0.855	0.975	0.3	0	0	0.42	0
9	−0.23	1.17	0	0	0.7	0.7	0
10	−0.23	1.17	0	0	0.7	0.7	0
11	−0.53	1.17	0	0.3	0.7	0.7	0
12	0.07	1.17	0.3	0	0.7	0.7	0
13	−0.565	1.365	0	0	1.4	1.4	0
14	−0.565	1.365	0	0	1.4	1.4	0
15	−0.565	1.365	0	0	1.4	1.4	0
16	−0.565	1.365	0	0	1.4	1.4	0
17	−0.17	1.95	0	0.3	0.84	0.98	0
18	−0.17	1.95	0	0.3	0.84	0.98	0
19	−0.17	1.95	0	0.3	0.84	0.98	0
20	−0.17	1.95	0	0.3	0.84	0.98	0
21	1.315	1.755	0	0.3	0	0.14	0
22	1.315	1.755	0	0.3	0	0.14	0
23	1.315	1.755	0	0.3	0	0.14	0
24	1.315	1.755	0	0.3	0	0.14	0

the available capacity in year 1 is not sufficient anymore for the flowed power in year 2. As a result, the line expansion is done in year 2 and the capacity is expanded. In all years, the capacity of the line is greater than the flowed power taking into account enough reserve margin.

TABLE 4.10   Exchanged power (per MW) with the grid in all years.

Hour	Year 1	Year 2	Year 3	Year 4	Year 5	Year 6
1	0.46	0.84	0.895	0.875	1.03	1.135
2	0.46	0.54	0.595	0.975	1.03	1.135
3	0.46	0.54	0.895	0.975	1.03	1.135
4	0.46	0.54	0.895	0.975	1.03	1.135
5	0.58	0.42	0.735	0.475	0.79	0.855
6	0.58	0.72	0.735	0.775	0.79	0.855
7	0.58	0.72	0.535	0.775	0.79	0.855
8	0.58	0.72	0.435	0.775	0.79	0.855
9	0.2	0.12	−0.29	−0.07	−0.18	−0.23
10	0	0.12	−0.29	−0.07	−0.48	−0.23
11	0.2	0.12	0.31	−0.07	0.12	−0.53
12	0.4	0.12	0.31	−0.07	−0.18	0.07
13	−0.3	−0.36	−0.655	−1.215	−1.21	−1.435
14	−0.1	−0.66	−0.655	−0.915	−1.21	−1.435
15	−0.1	−0.06	−0.655	−0.915	−1.21	−1.435
16	0.1	−0.36	−0.655	−0.615	−1.21	−1.435
17	0.28	0.12	0.01	−0.05	−0.16	−0.17
18	0.28	0.12	0.01	−0.05	−0.16	−0.17
19	0.28	0.12	0.01	−0.05	−0.16	−0.17
20	0.28	0.12	0.01	−0.05	−0.16	−0.17
21	0.8	1.02	0.835	1.295	1.11	1.315
22	0.8	1.02	1.135	0.995	1.11	1.315
23	0.8	1.02	1.135	0.995	1.11	1.315
24	0.8	1.02	1.035	1.095	1.11	1.315

The flowed power through the line at hour 15 and the line capacity is depicted in Fig. 4.7. It is clear that the expansion in year 2 is needed to handle the power growth. Similar results are presented for hour 21 in Fig. 4.8. In all hours, the capacity of the line is greater or equal to the flowed power through the line.

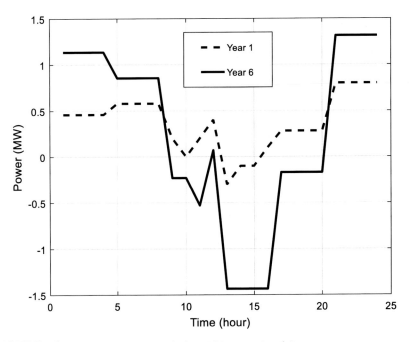

FIGURE 4.5 Exchanged power with the grid in years 1 and 6.

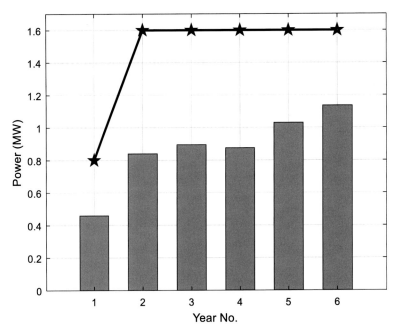

FIGURE 4.6 Power flow through line at hour 1 versus line capacity over the planning horizon. *Line*: capacity of line in each year. *Bars*: flowed power through line in each year.

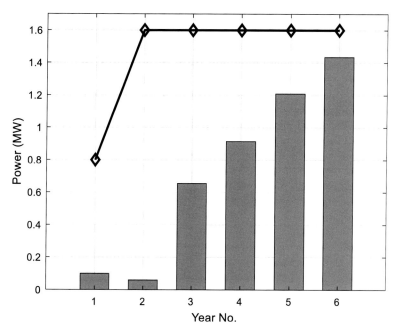

FIGURE 4.7 Power flow in the line at hour 15 versus line capacity over the planning horizon. *Line*: capacity of line in each year. *Bars*: flowed power through line in each year.

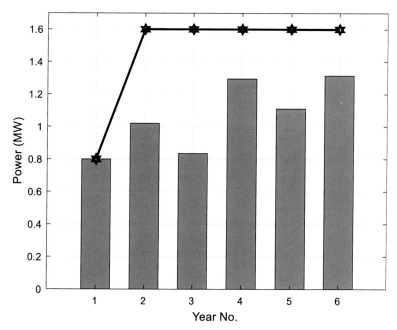

FIGURE 4.8 Power flow in the line at hour 21 versus line capacity over the planning horizon. *Line*: capacity of line in each year. *Bars*: flowed power through line in each year.

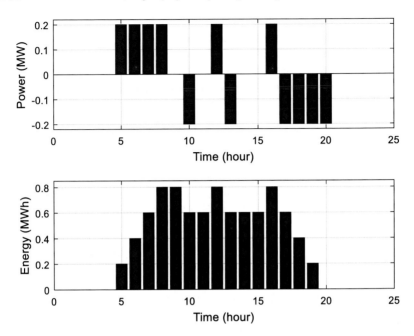

FIGURE 4.9   Battery operation and energy in year 1.

Figs. 4.9 and 4.10 demonstrate the battery charge-discharge operation and the battery energy in years 1 and 6, respectively. The battery often stores energy at the initial hours of the day between hours 1 and 10. It is afterward discharged at the last hours of the day between hours 17 and 23. Such energy arbitraging is efficient to reduce energy cost as well as deal with the shortage of renewable energy. The level of charging–discharging power and the stored energy are significantly increased from years 1 to 6. It is seen that the level of charging-discharging power in year 1 is about 0.2 MW but it is about 0.3 MW in year 6 because of installing new converters on the battery. As well, the maximum capacity of the battery in year 1 is 0.8 MWh while it is 2.4 MWh in year 6 because of expanding battery capacity.

4.6 Impacts of input data on planning

The input data such as technical parameters, prices, and costs make considerable impacts on the planning and outputs. The results of the planning wholly depend on the input data to the model and especially the prices. Considering the inaccurate input data leads to incorrect outputs or makes the results far from realistic. In order to study such points, the battery price is increased by 100% and the results of the planning are

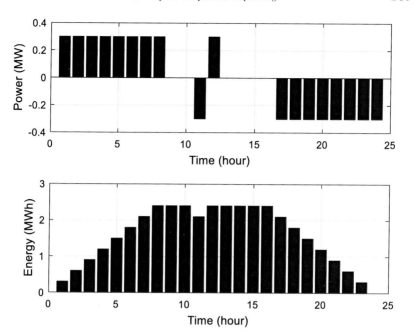

FIGURE 4.10   Battery operation and energy in year 6.

TABLE 4.11   Installed resources after increasing battery price by 100%.

Resource	Available resources in year 0	New installed resources					
		Year 1	Year 2	Year 3	Year 4	Year 5	Year 6
Solar system	1	1	1	1	1	1	1
Wind system	1	1	1	1	1	1	1
Line	1	0	1	0	0	0	0
Diesel generator	1	0	0	0	0	0	0
Power of battery	1	0	1	0	0	0	0
Capacity of battery	1	1	1	1	0	0	0

listed in Table 4.11. The results of the current case (increasing battery price by 100%) are compared to the original case (battery price is on the normal level). It is seen that the plan does not install many battery powers and capacities and the expansion of the battery is limited. The

TABLE 4.12   Planning cost after increasing battery price by 100%.

		Current case	Original case
Annual investment costs	Investment cost of solar system ($/year)	164,322.550	164,322.55
	Investment cost of wind system ($/year)	274,818.362	274,818.362
	Investment cost of line system ($/year)	11,745.650	11,745.65
	Investment cost of battery power ($/year)	15,346.440	15,730.524
	Investment cost of battery capacity ($/year)	46,076.544	36,626.724
	Investment cost of diesel generator ($/year)	0	0
Annual operational costs	Operational cost of solar system ($/year)	27,588.842	27,588.842
	Operational cost of wind system ($/year)	46,140.474	46,140.474
	Operational cost of line system ($/year)	2669.772	2669.772
	Operational cost of battery power ($/year)	3488.227	3891.092
	Operational cost of battery capacity ($/year)	10,503.090	6893.673
	Operational cost of diesel generator ($/year)	0	8343.024
	Cost of purchasing power from grid ($/year)	1,011,256.248	968,062.199
Total planning cost ($/year)		1,613,956.199	1,566,832.887

microgrid prefers to use other resources instead of installing new batteries. The planning costs are summarized in Table 4.12. For instance, the operational cost of the diesel generator is reduced and the microgrid purchases extra power from the external grid. The reason for installing and using the battery is to shift energy and reduce energy costs. When there is not enough battery in the microgrid, the shifted power is reduced and consequently the cost of purchasing power from the external grid is increased.

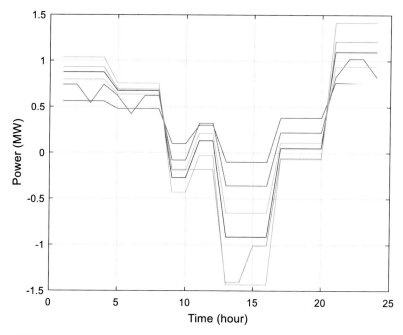

FIGURE 4.11   Traded power with the grid in all years.

Fig. 4.11 presents the traded power with the grid in all years and it is clear that the power patterns from years 1 to 6 are similar. The energy is not significantly shifted over the hours because the microgrid is not equipped with a large battery capacity.

If the microgrid is connected to the external grid via a longer line, the price of the line increases significantly. It is assumed that the line length increases by 300% resulting in 300% growth in the line price. In this case, the planning cost increases to 1,593,574.939 $/year, and the installed resources are presented in Table 4.13. The plan postpones the time of installing a new line from year 2 in the original case to year 3 in the current case. As well, the time of installing battery power changes together with the time of installing the line. The plan installs new lines in the further years in order to deal with the high price of lines. Since the costs are discounted back to the present year, installing lines in further years reduces the planning cost. As a result, the planning defers the line investment and deals with energy-power issues by using other resources rather than the line.

Table 4.14 presents the installed resources after increasing the battery operational cost to 15% and decreasing the battery lifespan to 2 years. In this case, the planning cost increases to 1,609,277.488 $/year. Since

TABLE 4.13   Installed resources after increasing line length by 300%.

Resource	Available resources in year 0	New installed resources					
		Year 1	Year 2	Year 3	Year 4	Year 5	Year 6
Solar system	1	1	1	1	1	1	1
Wind system	1	1	1	1	1	1	1
Line	1	0	0	0	1	0	0
Diesel generator	1	0	0	0	0	0	0
Power of battery	1	0	1	0	1	0	0
Capacity of battery	1	1	1	1	1	1	0

TABLE 4.14   Results after increasing battery cost to 15% and decreasing lifespan to 2 years.

Resource	Available resources in year 0	New installed resources					
		Year 1	Year 2	Year 3	Year 4	Year 5	Year 6
Solar system	1	1	1	1	1	1	1
Wind system	1	1	1	1	1	1	1
Line	1	0	1	0	0	0	0
Diesel generator	1	0	0	0	0	0	0
Power of battery	1	0	1	0	0	0	0
Capacity of battery	1	1	1	1	0	0	0

the battery lifetime is short and its operational cost is high, it is not an efficient resource for the microgrid. The plan reduces the installed batteries due to the inefficient operation of the battery.

Fig. 4.12 shows the battery operation in year 6 under the current case. The battery shows an operation similar to the previous cases because changing the battery capacity does not make effects on the operation pattern. Changing the battery capacity only varies the level of charged−discharged powers as well as the level of stored energy inside

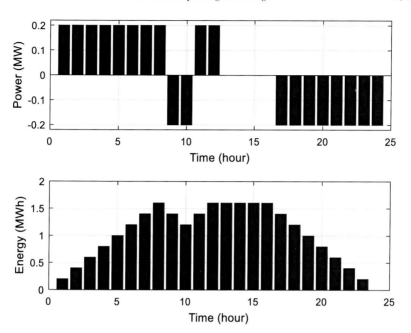

FIGURE 4.12   Battery operation in year 6 under the current case.

the battery. The battery regardless of its power and capacity is used to shift energy and such operation pattern is seen under small or large sizes of battery.

4.7 Resilient planning under outages

The outage of resources is a common event in microgrids. The outage of some resources may be more critical because of their role in the system. In the introduced microgrid, the line between the microgrid and the external network is very crucial, and disconnecting this line makes the microgrid islanded and off-grid. In such islanded state, the microgrid has to deal with the loads with its resources. As a result, the outage of the line may be regarded as one of the most critical outages in the microgrid. The planning considering the outage of the line at hours 17−22 is conducted again. In such modeling, the capacity of the line at hours 17−22 is set at zero which shows an off-grid operation during those hours.

Table 4.15 presents the installed resources under the outage of the line at hours 17−22. It is seen that plan installs two diesel generators in years 1 and 2. In the planning without the outage, the diesel generator

TABLE 4.15 Installed resources under outage of line at hours 17−22.

Resource	Available resources in year 0	New installed resources					
		Year 1	Year 2	Year 3	Year 4	Year 5	Year 6
Solar system	1	1	1	1	1	1	1
Wind system	1	1	1	1	1	1	1
Line	1	0	1	0	0	0	0
Diesel generator	1	1	1	0	0	0	0
Power of battery	1	0	1	1	1	1	0
Capacity of battery	1	1	1	1	1	1	0

is not installed but here the plan needs to utilize the diesel generators for handling the line outage. Generally, the diesel generator is a backup resource that is used in emergencies and this point is confirmed by the planning results. If the planning is performed on a microgrid that is faced with events and outages, the plan reinforces the microgrid with diesel generators. However, if the test microgrid is not faced with outages, the planning does not install new diesel generators and uses other resources. As a result, the outputs of the planning are completely associated with the input data to the model. Considering actual conditions in the model results in having an accurate output. Apart from the diesel generator, the plan also installs larger batteries on the microgrid to deal with line outages. During the outage of the line, the battery is needed to be operated as a resource for supplying loads. Such a point increases the requirement for installing extra batteries.

Table 4.16 presents the planning cost under the outage of the line at hours 17−22. In the original case, the investment cost of the diesel generator is zero while in the current case, it increases to 12,885.642 $/year. The operational cost of the diesel generator significantly increases from 8343.024 $/year to 459,709.698 $/year, because in the islanded condition the microgrid utilizes the diesel generator as much as needed. The cost of purchasing power from the grid reduces by about 45% because the connection to the grid is lost for many hours and the traded power is zero.

Table 4.17 lists the produced power by the diesel generator under outage of the line at hours 17−22. During the outage of the line, the diesel generator is operated by the microgrid to supply the loads. In years 4−6, the diesel is only operated for two hours 21 and 22 because the installed

TABLE 4.16 Planning cost under outage of the line at hours 17−22.

		Current case	Original case
Annual investment costs	Investment cost of solar system ($/year)	164,322.550	164,322.55
	Investment cost of wind system ($/year)	274,818.362	274,818.362
	Investment cost of line system ($/year)	11,745.650	11,745.65
	Investment cost of battery power ($/year)	28,569.420	15,730.524
	Investment cost of battery capacity ($/year)	36,626.724	36,626.724
	Investment cost of diesel generator ($/year)	12,885.642	0
Annual operational costs	Operational cost of solar system ($/year)	27,588.842	27,588.842
	Operational cost of wind system ($/year)	46,140.474	46,140.474
	Operational cost of line system ($/year)	2669.772	2669.772
	Operational cost of battery power ($/year)	4746.694	3891.092
	Operational cost of battery capacity ($/year)	6893.673	6893.673
	Operational cost of diesel generator ($/year)	459,709.698	8343.024
	Cost of purchasing power from grid ($/year)	561,333.587	968,062.199
Total planning cost ($/year)		1,638,051.089	1,566,832.887

batteries and solar-wind systems assist the microgrid to supply the loads, and the requirement for diesel generator power is reduced.

The capacity expansion should be proportional to the produced power by each resource. When the resources are required to produce extra power, their capacity should be expanded. Fig. 4.13 shows the diesel generator power at hour 21 versus the diesel generator capacity. It is seen that the capacity in the first year is not adequate for the produced power in the next years and the planning expands the diesel generator capacity in the second year. In all years, the produced power is less than the capacity.

TABLE 4.17   Power of diesel generator per MW under outage of the line at hours 17−22.

Hour	Year 1	Year 2	Year 3	Year 4	Year 5	Year 6
1−16	0	0	0	0	0	0
17	0.38	0.22	0.01	0	0	0
18	0.38	0.22	0.01	0	0	0
19	0.38	0.22	0.21	0	0	0
20	0.38	0.22	0.01	0	0	0
21	0.76	0.82	0.835	0.895	0.91	1.115
22	0.76	0.82	0.835	0.895	0.91	1.115
23	0	0	0	0	0	0
24	0	0	0	0	0	0

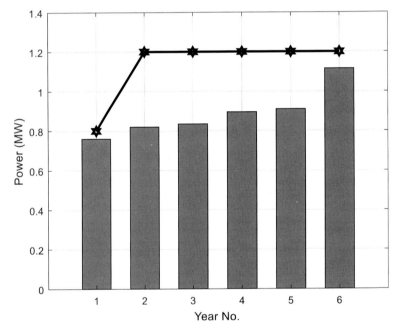

FIGURE 4.13   Diesel generator power at hour 21 versus diesel generator capacity. *Line*: capacity of diesel generator in each year. *Bars*: power of diesel generator in each year.

The traded power between the microgrid and the external network is listed in Table 4.18. It is clear that the exchanged power is zero under the outage of the line at hours 17−22. The microgrid is able to send power to

TABLE 4.18   Power to the grid per MW under outage of the line at hours 17−22.

Hour	Year 1	Year 2	Year 3	Year 4	Year 5	Year 6
1	0.56	0.74	0.595	1.075	1.23	0.835
2	0.56	0.74	0.595	0.675	0.73	0.835
3	0.56	0.74	0.895	1.075	0.73	1.335
4	0.56	0.74	0.695	0.675	1.23	0.835
5	0.48	0.62	0.735	0.875	0.49	0.955
6	0.48	0.42	0.735	0.875	0.89	1.055
7	0.48	0.62	0.735	0.875	0.99	1.055
8	0.48	0.42	0.735	0.475	0.99	1.055
9	0.2	0.12	−0.29	−0.47	−0.68	−0.23
10	0.2	0.12	0.01	0.33	0.32	−0.23
11	0.2	0.12	0.01	−0.07	−0.18	−0.23
12	0.2	0.12	0.31	−0.07	−0.18	−0.23
13	−0.2	−0.36	−0.955	−0.915	−1.21	−1.435
14	−0.2	−0.36	−0.655	−0.915	−1.21	−1.435
15	0	−0.56	−0.655	−0.915	−1.21	−1.435
16	0	−0.16	−0.355	−0.915	−1.21	−1.435
17	0	0	0	0	0	0
18	0	0	0	0	0	0
19	0	0	0	0	0	0
20	0	0	0	0	0	0
21	0	0	0	0	0	0
22	0	0	0	0	0	0
23	0.76	1.02	1.435	1.095	0.91	1.235
24	0.76	1.02	0.835	1.295	1.07	1.115

the external grid at many hours and this capability is increased from year 1 towards year 6, because the microgrid is equipped with extra resources in each year and the utmost resources are available in year 6.

The flowed power through the line at hour 3 and the line capacity in each year are shown in Fig. 4.14. The flowed power increases over the planning horizon years and the line capacity needs to be expanded. The line expansion is done in year 2 and the capacity is expanded. In all

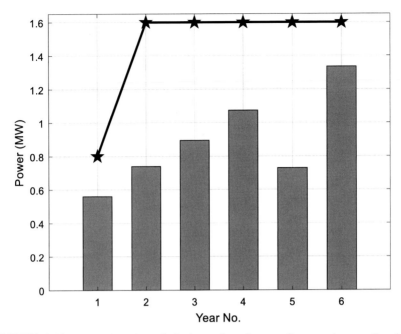

FIGURE 4.14 Power flow through the line at hour 3 versus line capacity over the planning horizon. *Line*: capacity of line in each year. *Bars*: flowed power through line in each year.

years, the capacity of the line is greater than the flowed power. Fig. 4.15 shows a similar output for hour 23. It is clear that the line capacity expansion is carried out in an appropriate year and the flowed power through the line is not limited because of line capacity restrictions.

Table 4.19 presents the powers of resources in year 6 under the outage of the line at hours 17–22. It is seen that the power equilibrium is passed at all hours. When the grid is disconnected, the resources like wind-solar systems and battery units operate to handle the load. At hours 21 and 22 when the solar energy becomes zero, the diesel generator is operated to supply the shortage of power.

Fig. 4.16 shows the power balance in the microgrid at some typical hours. At hour 3 the solar energy is zero and the battery and load consume energy. The wind system and the grid supply the required energy by 10% and 90%, respectively. At hour 19 the grid is disconnected and its power is set at zero. The load power is supplied by wind, solar, and battery discharge, where they produce 50%, 44%, and 6% of the required energy, respectively. At hour 21 the solar energy is zero and the grid is still disconnected. The wind, battery discharge, and diesel generator handle the load. Compared to hour 19, the produced power by wind increases, and the diesel generator is operated to compensate for the shortage of solar energy.

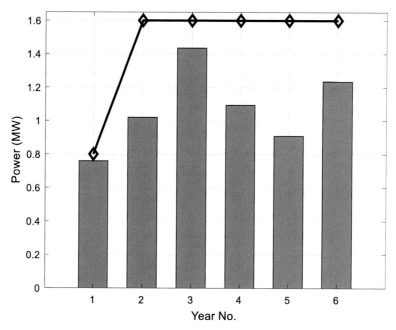

FIGURE 4.15   Power flow through the line at hour 23 versus line capacity over the planning horizon. *Line*: capacity of line in each year. *Bars*: flowed power through line in each year.

4.8 Planning without the energy storage system

The battery energy storage system plays a key role in the planning and shifts energy over the day hours to deal with both the shortage of energy and high-priced electricity. Eliminating the battery from planning could change the outputs substantially. In order to investigate this point, the planning without the battery energy storage system is simulated and the results are demonstrated in Table 4.20. It is seen that the system installs wind-solar units and lines in various years of the planning horizon. Table 4.21 presents the planning cost without the energy storage system. The investment and operational costs of battery power and capacity are zero. The operational cost of the diesel generator increases from 8343.024 to 10,385.746 $/year which shows a 25% growth. The cost of purchasing power from the grid also increases from 968,062.199 to 1,132,945.142 $/year which shows a 17% rise. Since there is no battery in the microgrid, it needs to receive more power from the external grid and the diesel generator. This point increases the operational cost of those resources.

The diesel generator is only operated in two years 1 and 6 at hours 21−24 as indicated by Fig. 4.17. The produced power is 60 kW in year 1 and 15 kW in year 6. As a result, the operating cost of the diesel

TABLE 4.19   Powers of resources in year 6 under outage of the line at hours 17–22.

Hour	Grid (MW)	Load (MW)	Battery charging (MW)	Battery discharging (MW)	Solar (MW)	Wind (MW)	Diesel generator (MW)
1	0.835	0.975	0	0	0	0.14	0
2	0.835	0.975	0	0	0	0.14	0
3	1.335	0.975	0.5	0	0	0.14	0
4	0.835	0.975	0	0	0	0.14	0
5	0.955	0.975	0.4	0	0	0.42	0
6	1.055	0.975	0.5	0	0	0.42	0
7	1.055	0.975	0.5	0	0	0.42	0
8	1.055	0.975	0.5	0	0	0.42	0
9	−0.23	1.17	0	0	0.7	0.7	0
10	−0.23	1.17	0	0	0.7	0.7	0
11	−0.23	1.17	0	0	0.7	0.7	0
12	−0.23	1.17	0	0	0.7	0.7	0
13	−0.565	1.365	0	0	1.4	1.4	0
14	−0.565	1.365	0	0	1.4	1.4	0
15	−0.565	1.365	0	0	1.4	1.4	0
16	−0.565	1.365	0	0	1.4	1.4	0
17	0	1.95	0	0.13	0.84	0.98	0
18	0	1.95	0	0.13	0.84	0.98	0
19	0	1.95	0	0.13	0.84	0.98	0
20	0	1.95	0	0.13	0.84	0.98	0
21	0	1.755	0	0.5	0	0.14	1.115
22	0	1.755	0	0.5	0	0.14	1.115
23	1.235	1.755	0	0.38	0	0.14	0
24	1.115	1.755	0	0.5	0	0.14	0

generator is not highly increased compared to the original case. The traded power with the upstream network is shown in Fig. 4.18. It is seen that both the received and sent powers increase from years 1 to 6. The new resources are installed every year and the microgrid comprises larger loads and resources in the final year. As a result, both the received power for loads and sent power from resources increase.

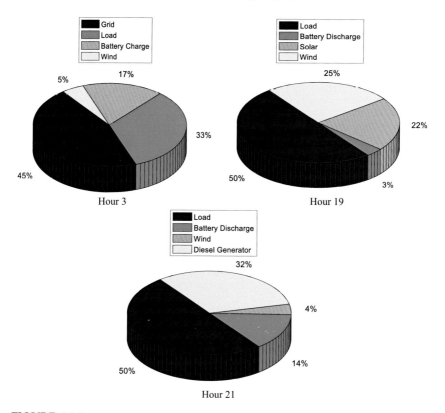

FIGURE 4.16 The generating-consuming power balance in a microgrid.

TABLE 4.20 Installed resources without the energy storage system.

Resource	Available resources in year 0	New installed resources					
		Year 1	Year 2	Year 3	Year 4	Year 5	Year 6
Solar system	1	1	1	1	1	1	1
Wind system	1	1	1	1	1	1	1
Line	1	0	1	0	0	0	0
Diesel generator	1	0	0	0	0	0	0
Power of battery	1	0	0	0	0	0	0
Capacity of battery	1	0	0	0	0	0	0

TABLE 4.21   Planning cost without the energy storage system.

		Current case	Original case
Annual investment costs	Investment cost of solar system ($/year)	164,322.550	164,322.55
	Investment cost of wind system ($/year)	274,818.362	274,818.362
	Investment cost of line system ($/year)	11,745.650	11,745.65
	Investment cost of battery power ($/year)	0.000	15,730.524
	Investment cost of battery capacity ($/year)	0.000	36,626.724
	Investment cost of diesel generator ($/year)	0.000	0.000
Annual operational costs	Operational cost of solar system ($/year)	27,588.842	27,588.842
	Operational cost of wind system ($/year)	46,140.474	46,140.474
	Operational cost of line system ($/year)	2669.772	2669.772
	Operational cost of battery power ($/year)	0.000	3891.092
	Operational cost of battery capacity ($/year)	0.000	6893.673
	Operational cost of diesel generator ($/year)	10,385.746	8343.024
	Cost of purchasing power from grid ($/year)	1,132,945.142	968,062.199
Total planning cost ($/year)		1,670,616.539	1,566,832.887

4.9 Stochastic expansion planning

In the stochastic modeling, five parameters of wind energy, solar energy, load power, electricity price, and fuel price are assumed as uncertain parameters. These parameters are modeled by a set of scenarios and the related equations need to be rewritten and defined over the set of scenarios. The purchased power from the upstream grid in all years of the planning horizon is reformulated as Eq. (4.42). It includes index "s" which shows scenarios of performance and this relationship is defined over the

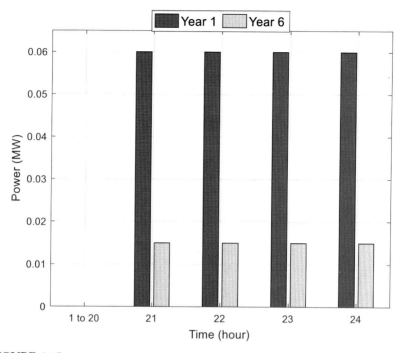

FIGURE 4.17   Produced power by a diesel generator in years 1 and 6 at hours 21–24.

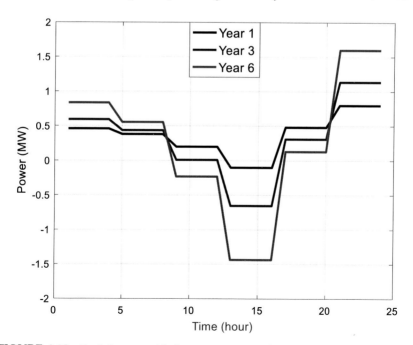

FIGURE 4.18   Traded power with the upstream network in years 1, 3, and 6.

set of scenarios of performance. The operational cost of the diesel generator in all years of the planning horizon is reformulated as Eq. (4.43). The exchanged power with the external grid is reformulated as Eq. (4.44). It is seen that the grid, load, solar, and wind powers are defined over a set of scenarios. The exchanged power with the grid must be less than the rated capacity of the line as presented by Eq. (4.45). The rest of the equations are similar to the deterministic model.

$$Zo_{grid} = \sum_{s=1}^{S} \left[\left\{ \sum_{y=1}^{Y} \left(\sum_{t=1}^{T} \left(P_{grid}^{t,y,s} \times F_{mk} \times C_{ele}^{t,y,s} \right) \times F_{dy} \times NPV \right) \right\} \times O_{pr}^{s} \right]$$

(4.42)

$$Zo_{diesel} = \sum_{s=1}^{S} \left[\left\{ \sum_{y=1}^{Y} \left(\sum_{t=1}^{T} \left(P_{diesel}^{t,y} \times F_{mk} \times C_{fuel}^{t,y,s} \right) \times F_{dy} \times NPV \right) \right\} \times O_{pr}^{s} \right]$$

(4.43)

$$P_{grid}^{t,y,s} = \left[P_{load}^{t,s} \times P_{rload}^{y} + P_{cb}^{t,y} \right] - \left[P_{db}^{t,y} + P_{solar}^{t,s} \times P_{rsolar}^{y} + P_{wind}^{t,s} \times P_{rwind}^{y} + P_{diesel}^{t,y} \right]$$
$$\forall t \in T$$
$$\forall y \in Y$$
$$\forall s \in S$$

(4.44)

$$\begin{cases} P_{grid}^{t,y,s} \leq + P_{rline}^{y} \\ P_{grid}^{t,y,s} \geq - P_{rline}^{y} \\ \forall y \in Y \\ \forall t \in T \\ \forall s \in S \end{cases}$$

(4.45)

It should be mentioned that the grid, load, solar, and wind powers are defined over a set of scenarios but the charge—discharge powers of the battery and the diesel generator power are modeled excluding scenarios and uncertainty. It means that the battery and diesel generator operations are constant under all scenarios of performance. They have a unique pattern for a 24-hour operation which is optimized to be feasible under all scenarios of performance.

4.9.1 Input data for the stochastic model

In the stochastic model, the fuel price, electricity price, load power, solar power, and wind power are assumed as uncertain parameters and each one is characterized by a set of scenarios.

Seven scenarios of performance are defined and the uncertain parameters under each scenario are presented here. Table 4.22 presents the fuel

TABLE 4.22 Fuel price under each scenario and probability of scenarios.

Scenario number	Fuel price ($/kWh)	Probability of occurrence for each scenario
1	0.10	0.22
2	0.09	0.18
3	0.12	0.16
4	0.08	0.14
5	0.13	0.12
6	0.14	0.10
7	0.15	0.08

price in each scenario as well as the probability of occurrence of each scenario is denoted. The electricity price under all scenarios of performance is listed in Table 4.23. The load, solar, and wind powers under all scenarios of performance are summarized in Tables 4.24–4.26, respectively.

The alterations of uncertain parameters under all scenarios of performance are shown in Fig. 4.19. It is seen that the uncertain area is covered by the introduced scenarios of performance. The fluctuations in the loads and prices may be greater than or less than the nominal values and this point is seen in the defined scenarios. On the other hand, the fluctuations of output powers in solar and wind systems cannot be greater than the rated power. As a result, all scenarios of performance comprise values less than the rated power.

4.9.2 Results of the stochastic planning

The proposed stochastic expansion planning is carried out on the given test case and the expanded resources are summarized in Table 4.27. Comparing this expansion planning to deterministic expansion planning demonstrates that stochastic expansion planning installs much more resources. The stochastic planning installs five solar systems, five wind units, two lines, four diesel generators, four battery converters, and five battery capacities during the planning horizon. Such large numbers of resources are applied to reinforce the microgrid for handling the uncertainty in the parameters. Since five parameters of the microgrid are modeled as uncertain parameters, the planning is faced with a wide range of uncertainty and it has to install many resources to cope with such extensive uncertainties. Table 4.28 reports the total number of available resources in each year over the planning horizon. It is seen that many resources are available in each year and about 33 resources are integrated into the microgrid in year 6.

TABLE 4.23   Electricity price in $/kWh under all scenarios of performance.

Hour	Scenario number						
	1	2	3	4	5	6	7
1	0.05	0.1	0.06	0.1	0.06	0.12	0.06
2	0.05	0.07	0.06	0.05	0.07	0.06	0.06
3	0.05	0.06	0.11	0.06	0.07	0.06	0.06
4	0.05	0.1	0.07	0.07	0.07	0.07	0.06
5	0.05	0.06	0.08	0.09	0.06	0.08	0.11
6	0.05	0.06	0.06	0.05	0.07	0.09	0.06
7	0.05	0.08	0.06	0.06	0.06	0.06	0.1
8	0.05	0.07	0.09	0.07	0.07	0.08	0.08
9	0.06	0.09	0.1	0.08	0.07	0.08	0.08
10	0.06	0.07	0.06	0.07	0.07	0.07	0.07
11	0.06	0.09	0.07	0.09	0.07	0.1	0.06
12	0.06	0.07	0.08	0.06	0.08	0.1	0.07
13	0.07	0.08	0.08	0.08	0.09	0.07	0.07
14	0.07	0.08	0.09	0.07	0.1	0.08	0.09
15	0.07	0.07	0.09	0.08	0.09	0.1	0.08
16	0.07	0.1	0.11	0.07	0.08	0.1	0.07
17	0.1	0.1	0.11	0.13	0.1	0.12	0.11
18	0.1	0.13	0.1	0.11	0.11	0.12	0.12
19	0.1	0.12	0.12	0.12	0.19	0.11	0.16
20	0.1	0.1	0.12	0.14	0.1	0.12	0.13
21	0.09	0.09	0.12	0.12	0.13	0.1	0.09
22	0.09	0.11	0.11	0.15	0.14	0.11	0.13
23	0.09	0.09	0.12	0.13	0.11	0.12	0.12
24	0.09	0.12	0.1	0.09	0.11	0.1	0.13

Table 4.29 signifies the investment and operational costs of stochastic programming. The results are compared to the deterministic model. The investment cost of the line increases from 11,745.65 to 23,519.790 $/year which shows about 50% augmentation. The investment cost of the battery power also raises from 29,998.314 to 15,730.524 $/year which shows

TABLE 4.24 Loading in percentage under all scenarios of performance.

Hour	Scenario number						
	1	2	3	4	5	6	7
1	50	47	58	62	79	68	80
2	50	48	61	80	20	74	99
3	50	52	21	76	66	30	57
4	50	59	59	47	71	82	73
5	50	78	31	31	51	67	48
6	50	72	65	41	61	53	72
7	50	55	69	39	75	58	60
8	50	55	57	46	69	56	51
9	60	72	61	71	85	72	96
10	60	51	76	30	73	88	53
11	60	82	102	76	52	77	59
12	60	57	84	49	60	63	70
13	70	81	70	92	80	78	74
14	70	95	77	70	83	82	94
15	70	97	68	71	86	90	103
16	70	92	78	58	59	98	64
17	100	111	126	92	100	92	100
18	100	111	132	106	157	115	111
19	100	90	78	102	101	109	97
20	100	108	120	105	142	127	132
21	90	115	82	114	117	92	116
22	90	124	93	107	92	91	107
23	90	111	113	119	95	99	106
24	90	101	107	100	94	59	99

about 50% growth. While the deterministic model does not install a diesel generator, the stochastic model invests 24,573.087 \$/year for installing new diesel generators. The operational cost of the diesel generators increases to 1,458,186.39 which demonstrates too much operation of those resources.

TABLE 4.25   Solar power in percentage under all scenarios of performance.

Hour	Scenario number						
	1	2	3	4	5	6	7
1	0	0	0	0	0	0	0
2	0	0	0	0	0	0	0
3	0	0	0	0	0	0	0
4	0	0	0	0	0	0	0
5	0	0	0	0	0	0	0
6	0	0	0	0	0	0	0
7	0	0	0	0	0	0	0
8	0	0	0	0	0	0	0
9	50	41	33	36	38	42	28
10	50	22	18	29	44	37	34
11	50	33	45	29	27	24	24
12	50	30	11	38	27	50	17
13	100	74	80	92	96	60	87
14	100	99	92	90	65	80	99
15	100	85	84	99	95	50	96
16	100	55	87	76	79	81	85
17	60	59	52	46	11	55	53
18	60	56	34	39	54	51	51
19	60	16	11	33	49	39	51
20	60	0	13	33	40	27	57
21	0	0	0	0	0	0	0
22	0	0	0	0	0	0	0
23	0	0	0	0	0	0	0
24	0	0	0	0	0	0	0

Table 4.30 presents the powers of all resources in the first year of the planning horizon under scenario 1. It is seen that the diesel generator is only operated at hours 18–20 and 22. During those hours, the grid power is negative which means injecting power into the grid by the microgrid. In other words, the microgrid uses a diesel generator for

TABLE 4.26 Wind power in percentage under all scenarios of performance.

Hour	Scenario number						
	1	2	3	4	5	6	7
1	10	0	0	0	0	6	0
2	10	1	0	0	7	0	0
3	10	0	0	0	1	0	0
4	10	0	0	7	0	0	0
5	30	0	0	0	0	0	0
6	30	29	10	17	26	8	21
7	30	19	2	13	22	8	22
8	30	23	6	13	22	22	24
9	50	48	16	42	14	33	43
10	50	32	33	46	17	29	33
11	50	27	39	21	38	50	30
12	50	17	46	47	11	44	30
13	100	62	91	74	74	84	90
14	100	87	49	97	85	51	66
15	100	92	91	84	92	78	100
16	100	77	95	99	95	98	95
17	70	39	47	65	48	56	60
18	70	69	50	48	0	58	53
19	70	67	55	69	57	58	64
20	70	63	27	49	61	64	65
21	10	0	0	6	0	0	9
22	10	0	0	0	0	0	0
23	10	3	0	0	0	0	0
24	10	0	0	0	0	0	0

producing power and sending it to the upstream network. Most of the battery discharge operations are seen at hours 17–20 when the micro-grid aims to send power to the external network. At some periods like hour 9, the grid power is zero which means the loads are supplied by internal resources.

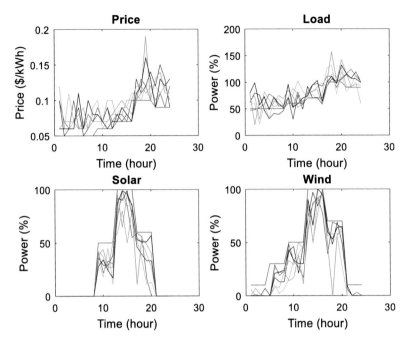

FIGURE 4.19   Uncertain parameters under all scenarios of performance.

TABLE 4.27   Number of installed resources in each year.

Resource	Available resources in year 0	New installed resources					
		Year 1	Year 2	Year 3	Year 4	Year 5	Year 6
Solar system	1	1	1	1	1	1	1
Wind system	1	1	1	1	1	1	1
Line	1	1	0	1	0	0	0
Diesel generator	1	1	1	1	1	0	0
Power of battery	1	1	1	1	1	0	0
Capacity of battery	1	1	1	1	1	1	0

Table 4.31 shows the powers of all resources in the last year of the planning horizon under scenario 1. It is seen that the powers of load and resources increase compared to the first year because of installing

TABLE 4.28 Total number of available resources in each year.

Resource	Year 0	Year 1	Year 2	Year 3	Year 4	Year 5	Year 6
Solar system	1	2	3	4	5	6	7
Wind system	1	2	3	4	5	6	7
Line	1	2	2	3	3	3	3
Diesel generator	1	2	3	4	5	5	5
Power of battery	1	2	3	4	5	5	5
Capacity of battery	1	2	3	4	5	6	6

TABLE 4.29 Investment-operational costs and total planning cost.

		Stochastic planning	Deterministic planning
Annual investment costs ($/year)	Investment cost of solar system ($/year)	164,322.550	164,322.55
	Investment cost of wind system ($/year)	274,818.362	274,818.362
	Investment cost of line system ($/year)	23,519.790	11,745.65
	Investment cost of battery power ($/year)	29,998.314	15,730.524
	Investment cost of battery capacity ($/year)	36,626.724	36,626.724
	Investment cost of diesel generator ($/year)	24,573.087	0
Annual operational costs ($/year)	Operational cost of solar system ($/year)	27,588.842	27,588.842
	Operational cost of wind system ($/year)	46,140.474	46,140.474
	Operational cost of line system ($/year)	5368.940	2669.772
	Operational cost of battery power ($/year)	6246.610	3891.092
	Operational cost of battery capacity ($/year)	6893.673	6893.673
	Operational cost of diesel generator ($/year)	1,458,186.399	8343.024
	Cost of purchasing power from grid ($/year)	962,650.545	968,062.199
Total planning cost ($/year)		3,066,934.311	1,566,832.887

TABLE 4.30   Powers of resources in year 1 of planning horizon under scenario 1.

Hour	Grid (MW)	Load (MW)	Battery charging (MW)	Battery discharging (MW)	Solar (MW)	Wind (MW)	Diesel generator (MW)
1	0.46	0.5	0	0	0	0.04	0
2	0.66	0.5	0.2	0	0	0.04	0
3	0.66	0.5	0.2	0	0	0.04	0
4	0.66	0.5	0.2	0	0	0.04	0
5	0.18	0.5	0	0.2	0	0.12	0
6	0.58	0.5	0.2	0	0	0.12	0
7	0.58	0.5	0.2	0	0	0.12	0
8	0.38	0.5	0	0	0	0.12	0
9	0	0.6	0	0.2	0.2	0.2	0
10	0.4	0.6	0.2	0	0.2	0.2	0
11	0	0.6	0	0.2	0.2	0.2	0
12	0.4	0.6	0.2	0	0.2	0.2	0
13	−0.1	0.7	0	0	0.4	0.4	0
14	−0.3	0.7	0	0.2	0.4	0.4	0
15	0.1	0.7	0.2	0	0.4	0.4	0
16	−0.1	0.7	0	0	0.4	0.4	0
17	0.28	1	0	0.2	0.24	0.28	0
18	−0.52	1	0	0.2	0.24	0.28	0.8
19	−0.52	1	0	0.2	0.24	0.28	0.8
20	−0.52	1	0	0.2	0.24	0.28	0.8
21	1.06	0.9	0.2	0	0	0.04	0
22	−0.14	0.9	0	0.2	0	0.04	0.8
23	0.86	0.9	0	0	0	0.04	0
24	0.86	0.9	0	0	0	0.04	0

new resources and expanding the capacities. The sent power to the grid by the microgrid increases because of the larger capacities of the microgrid in year 6.

Fig. 4.20 shows diesel generator operation in all years of the planning horizon under all scenarios of performance. In each year, the

TABLE 4.31   Powers of resources in year 6 of planning horizon under scenario 1.

Hour	Grid (MW)	Load (MW)	Battery charging (MW)	Battery discharging (MW)	Solar (MW)	Wind (MW)	Diesel generator (MW)
1	0.835	0.975	0	0	0	0.14	0
2	1.305	0.975	0.47	0	0	0.14	0
3	1.335	0.975	0.5	0	0	0.14	0
4	1.335	0.975	0.5	0	0	0.14	0
5	0.055	0.975	0	0.5	0	0.42	0
6	1.055	0.975	0.5	0	0	0.42	0
7	1.055	0.975	0.5	0	0	0.42	0
8	0.985	0.975	0.43	0	0	0.42	0
9	−0.73	1.17	0	0.5	0.7	0.7	0
10	0.27	1.17	0.5	0	0.7	0.7	0
11	−0.73	1.17	0	0.5	0.7	0.7	0
12	0.27	1.17	0.5	0	0.7	0.7	0
13	−1.435	1.365	0	0	1.4	1.4	0
14	−1.935	1.365	0	0.5	1.4	1.4	0
15	−0.935	1.365	0.5	0	1.4	1.4	0
16	−1.435	1.365	0	0	1.4	1.4	0
17	−0.37	1.95	0	0.5	0.84	0.98	0
18	−2.37	1.95	0	0.5	0.84	0.98	2
19	−2.37	1.95	0	0.5	0.84	0.98	2
20	−2.37	1.95	0	0.5	0.84	0.98	2
21	1.734	1.755	0.119	0	0	0.14	0
22	−0.885	1.755	0	0.5	0	0.14	2
23	1.596	1.755	0	0.019	0	0.14	0
24	1.615	1.755	0	0	0	0.14	0

diesel generator has a unique operating pattern under all scenarios of performance. In other words, the diesel generator operation does not change while the uncertain parameters vary from one value to another value. Such a robust operating pattern for the diesel generator is feasible under all scenarios of performance. In the first year, the

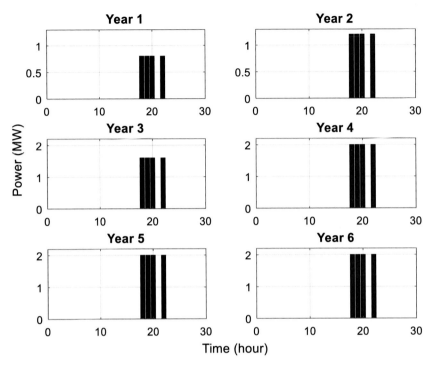

FIGURE 4.20   Diesel generator operation in all years under all scenarios of performance.

produced power is 0.8 MW and it is produced in a few hours. The produced power in the second year increases to 1.2 MW and it is 2 MW in year 6.

Figs. 4.21 and 4.22 show the battery charge-discharge and energy in years 1 and 6 under all scenarios of performance. Similar to the diesel generator, the battery has a unique operating pattern in each year under all scenarios of performance. In other words, a robust operation pattern is designed for operating in each year under all scenarios of performance. The battery is charged at many hours and discharged at many other hours because of uncertainty in the system.

Figs. 4.23 and 4.24 demonstrate the exchanged power with the external grid in years 1 and 6 under all scenarios of performance. It is seen that the exchanged power with the grid is altered from one scenario to another scenario. Unlike the battery and diesel generator, the exchanged power with the grid does not have a unique operating pattern under all scenarios of performance. In other words, the variations in the uncertain parameters make an impact on the power and change it. The uncertainty in the system is handled and managed by the exchanged power with the grid. When the power of load or renewables changes, such

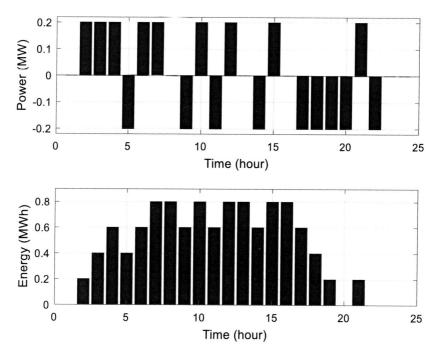

FIGURE 4.21 Battery charge–discharge and energy in year 1 under all scenarios of performance.

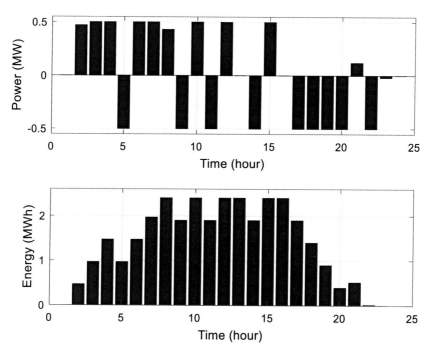

FIGURE 4.22 Battery charge–discharge and energy in year 6 under all scenarios of performance.

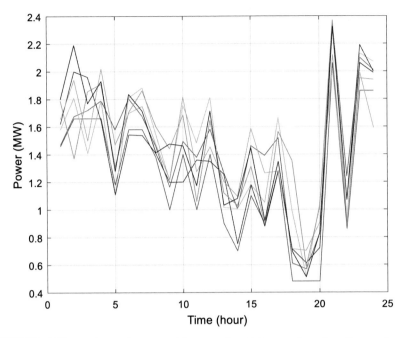

FIGURE 4.23   Exchanged power with external grid in year 1 under all scenarios of performance.

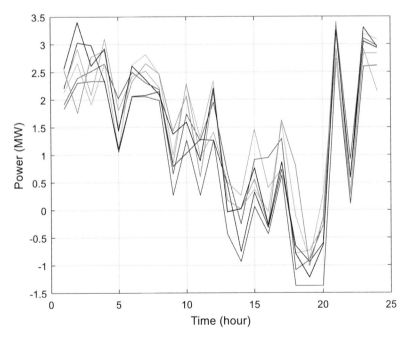

FIGURE 4.24   Exchanged power with external grid in year 6 under all scenarios of performance.

alterations are controlled by the grid power while the battery and the diesel generator are not responsible to handle such uncertainty.

Fig. 4.25 represents diesel generator power at hour 20 versus diesel generator capacity in each year. It is clear that the capacity and produced power both increase in years 1 to 4. In each year, the produced power is less than the capacity. The final capacity is 2 MW and all the capacity is used to produce power.

Tables 4.32 and 4.33 denote the powers at hours 20 and 21 in year 6 under all scenarios of performance. It is clear that the battery and the diesel generator show a unique operation under all scenarios. In Table 4.32, the battery discharged power is 0.5 MW and it is constant under all scenarios of performance. The diesel generator produces 2 MW under all scenarios of performance. The wind, solar, load, and grid powers change under scenarios and get different values from one scenario to another one. In Table 4.33, the diesel generator and battery charging powers are 0 and 0.119 MW under all scenarios of performance while the other powers change together with scenarios.

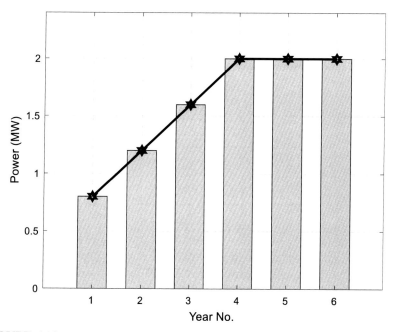

FIGURE 4.25 Diesel generator power at hour 20 versus its capacity. *Line*: capacity of diesel generator in each year. *Bars*: power of diesel generator in each year.

TABLE 4.32   Powers at hour 20 in year 6 under all scenarios of performance.

Scenario no.	Grid (MW)	Load (MW)	Battery charging (MW)	Battery discharging (MW)	Solar (MW)	Wind (MW)	Diesel generator (MW)
1	−2.37	1.95	0	0.5	0.84	0.98	2
2	−1.276	2.106	0	0.5	0	0.882	2
3	−0.72	2.34	0	0.5	0.182	0.378	2
4	−1.601	2.048	0	0.5	0.462	0.686	2
5	−1.145	2.769	0	0.5	0.56	0.854	2
6	−1.298	2.477	0	0.5	0.378	0.896	2
7	−1.634	2.574	0	0.5	0.798	0.91	2

TABLE 4.33   Powers at hour 21 in year 6 under all scenarios of performance.

Scenario no.	Grid (MW)	Load (MW)	Battery charging (MW)	Battery discharging (MW)	Solar (MW)	Wind (MW)	Diesel generator (MW)
1	1.734	1.755	0.119	0	0	0.14	0
2	2.361	2.242	0.119	0	0	0	0
3	1.718	1.599	0.119	0	0	0	0
4	2.258	2.223	0.119	0	0	0.084	0
5	2.4	2.282	0.119	0	0	0	0
6	1.913	1.794	0.119	0	0	0	0
7	2.255	2.262	0.119	0	0	0.126	0

4.10 Conclusions

In this chapter, long-term expansion planning was addressed on the residential microgrid. The objective of the plan was to minimize the annualized planning cost including the annual operational cost plus the annual investment cost. The microgrid was integrated with six elements including wind, solar, battery storage, battery converter, diesel generator, and external grid. All of these elements were expanded to deal with load growth during a 6-year planning horizon. The results showed that the planning installed 10 wind and solar systems on the microgrid because of their lower investment and operational costs. The diesel generator was not installed by the planning because of the high operating cost. The diesel generator is more useful when the microgrid faces outages. Since the outages and events are not modeled here, the planning afterward does not install the diesel generator. The planning also installed one line, two battery converters, and four new batteries in the microgrid over the planning horizon. The power balance in the microgrid was passed at all hours in all years. The load and the battery charging power were the consumers of energy. The discharging power of the battery, wind, solar, and diesel units were the power generators, and the grid power operated on both consuming and generating states. The battery often stored energy at the initial hours of the day between hours 1 and 10. It was afterward discharged at the last hours of the day between hours 17 and 23. In the first year, the microgrid sent power to the upstream grid at hours 13−15, but in year 6, the microgrid was able to send power to the grid in extra hours such as hours 9−20.

When the battery price increased by 100%, the plan did not install many batteries and the planning cost increased by 3%. Growing line length by 300% increased the planning cost to 1,593,574.939 $/year and decreasing battery lifespan to 2 years increased the planning cost to 1,609,277.488 $/year.

Under the outage of the line at hours 17−22, the plan installed two diesel generators in the microgrid in years 1 and 2, and the planning cost increased to 12,885.642 $/year. The operational cost of the diesel generator increased from 8343.024 to 459,709.698 $/year and the cost of purchasing power from the grid was reduced by 45%. At the hours when the grid was disconnected, the load power was successfully supplied by wind, solar, battery, and diesel generator. The planning without battery showed that the operational cost of the diesel generator increased by 25% and the cost of purchasing power from the grid increased by 17%.

In the stochastic modeling, the wind energy, solar energy, load power, electricity price, and fuel price were modeled by a set of scenarios. The stochastic expansion planning installed five solar and five wind units, two lines, four diesel generators, four battery converters, and five battery capacities during the planning horizon. It increased the planning cost by about 50%, whereas both the investment and operational costs were increased by 50%. The diesel generator and battery showed a unique operating pattern under all scenarios of performance, however, the exchanged power with the grid was altered from one scenario to another scenario to handle the introduced uncertainty.

References

[1] Hemmati R, Hooshmand R-A, Khodabakhshian A. Comprehensive review of generation and transmission expansion planning. IET Generation, Transmission & Distribution 2013;7(9):955−64.

[2] Saboori H, Hemmati R. Considering carbon capture and storage in electricity generation expansion planning. Ieee Transactions on Sustainable Energy 2016;7(4):1371−8.

[3] de Araujo RA, Torres SP, Filho JP, Castro CA, Van Hertem D. Unified AC transmission expansion planning formulation incorporating VSC-MTDC, FACTS devices, and reactive power compensation. Electric Power Systems Research 2023;216:109017.

[4] Hemmati R, Hooshmand R-A, Khodabakhshian A. State-of-the-art of transmission expansion planning: comprehensive review. Renewable and Sustainable Energy Reviews 2013;23:312−19.

[5] Mehrjerdi H, Hemmati R. Wind-hydrogen storage in distribution network expansion planning considering investment deferral and uncertainty. Sustainable Energy Technologies and Assessments 2020;39:100687.

[6] Saberi R, Falaghi H, Esmaeeli M, Ramezani M, Ashoornezhad A, Izadpanah R. Power distribution network expansion planning to improve resilience. IET Generation, Transmission & Distribution; n/a.

[7] Khob SAE, Moazzami M, Hemmati R. Advanced model for joint generation and transmission expansion planning including reactive power and security constraints of the network integrated with wind turbine. International Transactions on Electrical Energy Systems 2019;29(4).

[8] Hemmati R, Hooshmand R-A, Khodabakhshian A. Coordinated generation and transmission expansion planning in deregulated electricity market considering wind farms. Renewable Energy 2016;85:620−30.

[9] Zhong H, Zhang G, Tan Z, Ruan G, Wang X. Hierarchical collaborative expansion planning for transmission and distribution networks considering transmission cost allocation. Applied Energy 2022;307:118147.

[10] Saboori H, Hemmati R, Abbasi V. Multistage distribution network expansion planning considering the emerging energy storage systems. Energy conversion and management 2015;105:938−45.

[11] Wu T, Wei X, Zhang X, Wang G, Qiu J, Xia S. Carbon-oriented expansion planning of integrated electricity-natural gas systems with EV fast-charging stations. IEEE Transactions on Transportation Electrification 2022;8(2):2797−809.

[12] Franken M, Barrios H, Schrief AB, Moser A. Transmission expansion planning via power flow controlling technologies. IET Generation, Transmission & Distribution 2020;14(17):3530−8.

[13] Mehrjerdi H, Hemmati R, Mahdavi S, Shafie-khah M, Catalao JP. Multi-carrier microgrid operation model using stochastic mixed integer linear programming. IEEE Transactions on Industrial Informatics 2022;18(7):4674−87.

[14] Hemmati R, Saboori H, Siano P. Coordinated short-term scheduling and long-term expansion planning in microgrids incorporating renewable energy resources and energy storage systems. Energy. 2017;134:699−708.

[15] Mishra S, Bordin C, Tomasgard A, Palu I. A multi-agent system approach for optimal microgrid expansion planning under uncertainty. International Journal of Electrical Power & Energy Systems 2019;109:696−709.

[16] Saboori H, Mehrjerdi H, Jadid S. Mobile battery storage modeling and normal-emergency operation in coupled distribution-transportation networks. IEEE Transactions on Sustainable Energy 2022;13(4):2226−38.

District energy systems

© 2024 Elsevier Inc. All rights reserved.

Nomenclature

Indexes and sets

h	Index of homes
H	Set of homes in district energy system
s	Index of scenarios
S	Set of scenarios in stochastic model
t	Index of hours
T	Set of hours in one day

Parameters and variables

$BigM$	A very big number (1e6 in the presented examples)
C_{cost}	Penalty cost for curtailed loads (\$/kWh)
DG_{dot}	Diesel generator operating hours in one day
E_b^t	Energy of battery (kWh)
E_b^0	Initial energy inside battery (kWh)
E_{price}^t	Electricity price (\$/kWh)
E_{rb}	Rated capacity of battery (kWh)
F_{loss}	Factor showing energy loss (%)
F_{price}	Fuel price (\$/kWh)
$k_{lc}^{t,h}$	Load curtailment factor [0,1]
$k_{lc}^{t,h,s}$	Load curtailment factor in stochastic model [0,1]
k	Consecutive operating hours allowed for diesel generator
$L_c^{t,h}$	Curtailed load (kW)
P_{12}^t	Power from node 1 (bus 1) to node 2 (bus 2) (kW)
$P_{12}^{t,s}$	Power from node 1 (bus 1) to node 2 (bus 2) in stochastic model (kW)
P_{23}^t	Power from node 2 (bus 2) to node 3 (bus 3) (kW)
P_{34}^t	Power from node 3 (bus 3) to node 4 (bus 4) (kW)
P_{45}^t	Power from node 4 (bus 4) to node 5 (bus 5) (kW)
P_{cb}^t	Charged power to battery (kW)
P_{cgs}^t	Power of central generating system (kW)
$P_{crload}^{t,h}$	Power of critical loads (kW)
P_{dg}^t	Power of diesel generator (kW)
P_{db}^t	Discharged power from battery (kW)
P_{grid}^t	Power from upstream grid toward district (kW)
$P_{grid}^{t,s}$	Power from upstream grid toward district in stochastic model (kW)
P_{h45p}^t	Power from home 4 to home 5 in positive direction (kW)
P_{h45n}^t	Power from home 4 to home 5 in negative direction (kW)
P_{h54p}^t	Power from home 5 to home 4 in positive direction (kW)
P_{h54n}^t	Power from home 5 to home 4 in negative direction (kW)
$P_{home}^{t,h}$	Power from district to homes (kW)
$P_{home}^{t,h,s}$	Power from district to homes in stochastic model (kW)
$P_i^{t,s}$	Flowed power in each sector of district (kW)
$P_{load}^{t,h}$	Power of load (kW)
$P_{load}^{t,h,s}$	Power of load in stochastic model (kW)
P_{rb}	Rated power of battery charger (kW)
P_{rdg}	Rated power of diesel generator (kW)
P_{rgrid}	Rated capacity of transmission line in the district energy system (kW)
P_{rhome}^{lt}	Rated capacity of the line between home and district energy system (kW)

$P_{solar}^{t,h}$	Power of solar panels (kW)
$P_{solar}^{t,h,s}$	Power of solar panels in the stochastic model (kW)
$P_{wind}^{t,h}$	Power of wind turbine (kW)
$P_{wind}^{t,h,s}$	Power of wind turbine in the stochastic model (kW)
u_{cb}^{t}	Binary variable showing charging state of the battery
u_{db}^{t}	Binary variable showing discharging state of battery
u_{dg}^{t}	Binary variable showing diesel generator operation
u_{h45}^{t}	Binary variable showing power from home 4 to 5
u_{h54}^{t}	Binary variable showing power from home 5 to 4
Z_{dg}	Daily operating cost of diesel generator ($/day)
Z_{final}	Daily energy cost of the district energy system ($/day)
Z_{grid}	Daily energy purchased from the upstream grid ($/day)
Z_{lc}	Daily penalty cost for load curtailment ($/day)
Z_{p2p}	Daily cost of energy loss in peer-to-peer energy exchange ($/day)
η_b	Efficiency of battery (%)

5.1 Introduction

District energy systems (DESs) are characterized by several central resources that provide electricity, heating, cooling, or any combination of them for nearby buildings. The DESs are able to serve a variety of end-users such as residential districts, university campuses, hospital districts, airport zones, military bases, industrial districts, and central business districts [1]. In the DESs, the individual users may also be equipped with renewable or nonrenewable energy resources and produce energy. Such users are called prosumers. The prosumers produce energy, either for self-consumption or for consumption by other users. The prosumers may obtain implicit or explicit incentives for participating in the energy management system of the district. The prosumers are equipped with various types of resources like wind turbines, solar panels, battery energy storage systems, heat or cold storage devices, heating systems, and cooling systems, or even they would participate in the demand response program [2]. The energy management system of the district collects the data of the system including produced power of prosumers, loads of users, exchanged power with the external grid, and exchanged power with central generating resources. The optimal operating condition is afterward optimized for each user and prosumer in order to minimize the costs or maximize the individuals' profits. In the United States, the DESs are usually placed on university or hospital campuses, military bases, airport areas, and central business districts. Many cities such as New York, Boston, Philadelphia, San Francisco, and Denver have downtown DESs. The DES may be integrated with individual buildings and these buildings may be commonly owned or may have separate owners. Based on the US energy information administration report, about 660 DESs are working in the US [1].

The DESs provide benefits for both the users and the communities. As well, the external electrical grid benefits from DESs [3]. The benefits to users may be listed as higher energy efficiency, lower investment cost on energy equipment, no need for maintenance of energy resources, reduced noises-vibrations, increased well-being, and increased reliability because the industrial grade energy system is more robust than the commercial energy system installed at the building level. The benefits to the cities and communities may be summarized as reducing investment costs and quick return on the investment. The external electrical grid can also be benefited from DESs through collaborative operating such as peak-cutting, increasing the resilience of the system following outages and events, and increasing the reliability of the grid by supplying the loads from two directions of the grid and the DES [1,3]. The DESs make positive impacts on the environment by reducing greenhouse gases and increasing the adoption of renewable energies and electric vehicles.

The resilience of the energy system is one of the points that could be improved by the development of DESs [4]. Resilience is defined as the capacity to withstand or to recover rapidly from problems and setbacks, adapt to changes, and continue operating in the face of difficulty. In energy systems, resilience is the ability to prepare for and adapt to changing conditions and withstand and recover rapidly from disruptions. Resilience includes the ability to withstand and recover from deliberate attacks, accidents, or naturally occurring threats [5]. In the electrical networks, resilience has been investigated in various sectors from the generation sector to the end-user sector including electric power systems [6], distribution grids and microgrids [7], and home energy management systems [8]. Recently, emerging technologies are used for resilience improvement. For instance, the application of mobile energy storage systems for improving electrical grid resilience is broadly addressed [9,10]. Electric vehicles are managed to enhance resilience in urban energy systems as well [11]. The multicarrier energy systems and energy hubs are the other emerging technologies to improve resilience in electrical energy systems [12].

The DESs can also be modeled to deal with outages and events in order to improve the resilience of urban energy systems [4]. In the DESs, the local resources can supply the local loads when the upstream grid has faced an outage. During events, the DESs can recover some or all loads and reduce the unsupplied energy of loads [4]. As well, the DESs enable peer-to-peer (p2p) energy trading between users and such ability assists the community to deal with events. When one or some of the consumers are disconnected from the main grid, the adjacent prosumers can supply their loads by p2p energy exchange [13]. In the DESs, the prosumers and consumers should operate based on the optimal operating pattern. As well, the central resources owned by the DES are required to be optimized and operated. The exchanged energy between the DES and the upstream grid also needs to be optimized and scheduled. All of these items are modeled through one optimization programming and the optimal operating condition is achieved for all participants in the DES. The developed model may consider the outages and events for having a resilient DES. In such a resilient model, the system must be able to deal with internal and external outages.

In this chapter, energy management in DES is addressed. The details and highlights of this chapter are as follows:

- The DES is connected to the external electrical grid. The optimal energy exchange between the DES and the external grid is scheduled by the energy management system. The DES can take energy from the grid and send energy to the grid. The objective function of the energy management system is to minimize the daily energy cost in the DES.
- The developed DES is formed by six homes (or buildings) and each building comprises its own loads and renewable resources. Homes 1 and 3 are equipped with solar panels and home 5 is supplied with a wind turbine. These homes operate as prosumers. The other homes are consumers. The energy management system harvests maximum power from renewable resources in the homes.
- Each home is integrated with critical and noncritical loads. The critical loads must be supplied under any situation but the noncritical loads may be regulated, curtailed, or interrupted by DES. The optimal operating pattern of each load is scheduled by DES.
- The DES is connected to a central generating system (CGS). The CGS is formed by the diesel generator and battery energy storage system. The operating pattern of the diesel generator and the charging scheduling of the battery are optimized by DES. As well, the energy exchange between CGS and DES is optimally scheduled.
- The p2p energy exchange is adopted between homes 4 and 5 in the district. The optimal operation for p2p energy exchange is scheduled by DES.
- Several situations such as normal conditions, outage of lines, outage of homes, and off-grid operations are modeled, formulated, simulated, and discussed.
- The parametric uncertainty is taken into account and mixed integer stochastic programming is carried out to deal with uncertainty. Wind energy, solar energy, and load power are assumed to be the sources of uncertainty and they are modeled by a set of scenarios of performance.
- All the models are simulated in GAMS software and the practical examples with numerical data are presented.

5.1.1 Some research gaps in district energy systems

The main research gaps in DESs may be summarized as follows.

5.1.1.1 *User satisfaction and the topics from the point of view of users and building owners*

The well-being and comfort of end-users and building owners are very important and essential. It would be needed to define proper incentives and economic motivations for the individuals for participating in the energy management of DESs. The regulations need to be developed. The user satisfaction index is required to be regarded in the programming.

5.1.1.2 *Physical and cyber-attacks*

The DESs are developing rapidly. These systems operate a large number of facilities in a wide area including electrical transmission systems, thermal transmission systems, gas pipelines, water pipelines, control systems, and so forth. All of those resources need to be controlled and managed correspondingly and in real-time. The physical devices would face the physical attacks and the transferred data would be exposed to the cyber-attacks. A resilient and self-heading energy management system is needed for reliable and stable operation amid those challenges.

5.2 Developed district energy system

The developed DES is shown in Fig. 5.1. The DES is connected to the external electrical grid. The DES is formed by six individual homes (or buildings). Homes 1 and 3 are prosumers and equipped with solar panels as well as home 5 is supplied by a wind turbine. The other homes are consumers. A CGS is designed for DES. The CGS is integrated with the diesel generator and battery energy storage system. The DES is supplied from both directions including the grid and CGS. When one of these two resources is not available, the DES is able to supply the homes through another resource [4]. The p2p energy exchange is adopted between homes 4 and 5 [13].

Fig. 5.2 depicts the single-line diagram of DES. Every home is supplying two loads including the critical and noncritical loads. The critical loads are the important loads that cannot be curtailed or interrupted and they must be supplied at all hours. On the other hand, some loads are noncritical and they would be curtailed or interrupted by DES if necessary. When the DES faces a shortage of energy (e.g., outage of elements), it is allowed to manage the noncritical loads and switch them

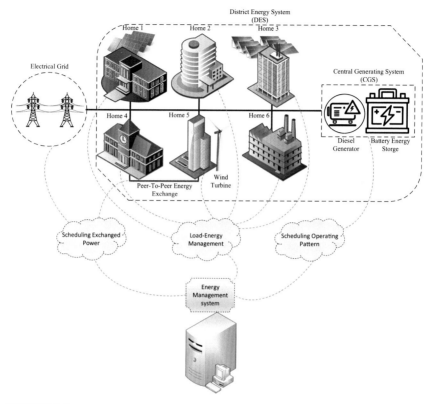

FIGURE 5.1 Developed district energy system.

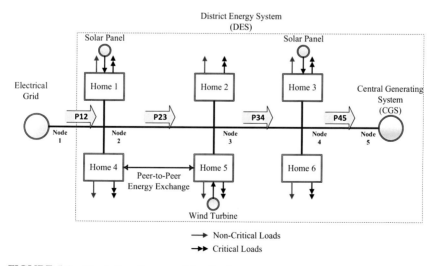

FIGURE 5.2 Single-line diagram of the district energy system.

off or regulate their power. The resources in the CGS including diesel generator and battery energy storage need to be optimally scheduled for operating in 24 hours. The produced power by the diesel generator at each hour and the charging–discharging scheduling of the battery at each hour should be determined.

The DES is modeled with four sectors between five nodes. The power from the grid to DES is defined by P_{12} and the power from CGS to DES is defined by P_{45}. The powers between sectors, P_{12}, P_{23}, P_{34}, and P_{45}, are defined as the free variables for which both positive and negative powers are acceptable. In other words, the direction of power can be changed from DES toward the grid.

Two homes 4 and 5 are modeled to exchange energy based on the p2p operation. In the p2p energy exchange, the homes can trade energy with each other when they are not connected to the DES or when the system faces events and outages. The energy management system needs to determine the following items:

- Optimal operating scheduling for resources in the CGS;
- Optimal load management in each individual home;
- Optimal energy exchange between DES and external grid;
- Optimal operating pattern for p2p energy exchange;
- Supplying critical loads under all events and outages;
- Harvesting maximum energy from renewable resources in individual homes;
- Optimal energy exchange between DES and prosumers.

5.3 Formulation of district energy system

The first cost in the DES is related to the purchased power from the external grid which is expressed by Eq. (5.1). The cost of consumed fuel for diesel generator operation is calculated by Eq. (5.2). The fuel cost for the diesel generator operation is usually more expensive than buying electricity from the external grid. Therefore, the DES uses the diesel generator only at the hours when the grid is not able to supply the loads. If the internal resources and the external grid are not able to handle the loads, the DES carries out the load management program and curtails some noncritical loads while all the critical loads are assured to be supplied. The penalty cost for load curtailment is expressed by Eq. (5.3). The p2p energy exchange between homes comes with some energy losses because the supplied energy to the home is taken from the adjacent home instead of the external grid and such point increases the route of power resulting in more energy losses. The cost of energy losses for p2p energy exchange is defined by Eq. (5.4). Finally, the daily energy cost of DES is achieved by Eq. (5.5). This cost is assumed as the objective function of optimization programming and needs to be minimized.

$$Z_{grid} = \sum_{t=1}^{T} \left(P_{grid}^t \times E_{price}^t \right) \tag{5.1}$$

$$Z_{dg} = \left(\sum_{t=1}^{T} \left(P_{dg}^t \right) \right) \times F_{price} \tag{5.2}$$

$$Z_{lc} = \left(\sum_{h=1}^{H} \left[\sum_{t=1}^{T} \left(L_c^{t,h} \right) \right] \right) \times C_{cost} \tag{5.3}$$

$$Z_{p2p} = \left(\sum_{t=1}^{T} \left(P_{h45p}^t + P_{h54p}^t \right) \times F_{loss} \times E_{price}^t \right) \tag{5.4}$$

$$Z_{final} = Z_{grid} + Z_{dg} + Z_{lc} + Z_{p2p} \tag{5.5}$$

The power from node 1 to node 2 is modeled as the received power from the grid and this point is indicated by Eq. (5.6). The power from node 1 to node 2 is used to supply homes 1 and 4 and the rest of this power is transferred to the next sector of DES. The power balance in nodes 2 to 5 is modeled through Eqs. (5.7) to (5.9). The flowed power between DES and CGS is expressed as Eq. (5.10).

$$P_{grid}^t = P_{12}^t \atop \forall t \in T \tag{5.6}$$

$$P_{12}^t = P_{23}^t + P_{home}^{t,h1} + P_{home}^{t,h4} \atop \forall t \in T \tag{5.7}$$

$$P_{23}^t = P_{34}^t + P_{home}^{t,h2} + P_{home}^{t,h5} \atop \forall t \in T \tag{5.8}$$

$$P_{34}^t = P_{45}^t + P_{home}^{t,h3} + P_{home}^{t,h6} \atop \forall t \in T \tag{5.9}$$

$$P_{45}^t = P_{cgs}^t \atop \forall t \in T \tag{5.10}$$

The main line between the upstream grid and the DES is continued toward CGS. The capacity of this line in all four sectors is equal and this point is modeled by Eqs. (5.11) to (5.14).

$$\begin{cases} P_{12}^t \leq P_{rgrid} \\ P_{12}^t \geq -P_{rgrid} \end{cases} \forall t \in T \tag{5.11}$$

$$\begin{cases} P^t_{23} \leq P_{rgid} \\ P^t_{23} \geq -P_{rgrid} \end{cases} \forall t \in T \tag{5.12}$$

$$\begin{cases} P^t_{34} \leq P_{rgrid} \\ P^t_{34} \geq -P_{rgrid} \end{cases} \forall t \in T \tag{5.13}$$

$$\begin{cases} P^t_{45} \leq P_{rgrid} \\ P^t_{45} \geq -P_{rgrid} \end{cases} \forall t \in T \tag{5.14}$$

The power from DES to homes is modeled by Eq. (5.15). Some homes are equipped with renewable resources such as wind turbines and solar panels and these resources are also modeled [7].

$$P^{t,h}_{home} = P^{t,h}_{load} \times k^{t,h}_{lc} - P^{t,h}_{solar} - P^{t,h}_{wind} \\ \forall t \in T \\ \forall h \in H, h \notin [h_4, h_5] \tag{5.15}$$

Home 4 and 5 are modeled to have p2p energy exchange. In order to model the energy losses in p2p operation, the flowed power between these two homes through p2p connection is calculated through Eqs. (5.16) to (5.23) based on a linear model. The power balance in home 4 is expressed by Eq. (5.16), where, the power from home 4 to home 5 is characterized by two terms of positive and negative powers. The positive term shows power from home 4 to home 5 and the negative term indicates the power from home 5 to home 4. A similar relationship is defined for home 5 by Eq. (5.17). The positive power from 4 to home 5 is equal to the negative power from 5 to home 4 and vice-versa as modeled by Eq. (5.18). The binary variables are defined in Eq. (5.19) to confirm the operation only in one direction, either from home 4 to 5 or from home 5 to 4. According to constraints Eqs. (5.20) to (5.23), the flowed power is calculated and it is confirmed to be only in one direction. The flowed power is used to calculate the energy losses in p2p energy exchange.

$$P^{t,h}_{home} = P^{t,h}_{load} \times k^{t,h}_{lc} - P^{t,h}_{solar} - P^{t,h}_{wind} + P^t_{h45p} - P^t_{h45n} \\ \forall t \in T \\ \forall h = [h_4] \tag{5.16}$$

$$P^{t,h}_{home} = P^{t,h}_{load} \times k^{t,h}_{lc} - P^{t,h}_{solar} - P^{t,h}_{wind} + P^t_{h54p} - P^t_{h54n} \\ \forall t \in T \\ \forall h = [h_5] \tag{5.17}$$

$$\begin{cases} P^t_{h54p} = P^t_{h45n} \\ P^t_{h45p} = P^t_{h54n} \end{cases} \forall t \in T \tag{5.18}$$

$$u^t_{h45} + u^t_{h54} \leq 1 \\ \forall t \in T \tag{5.19}$$

$$P^t_{h45p} \leq u^t_{h45} \times BigM$$
$$\forall t \in T$$

(5.20)

$$P^t_{h54n} \leq u^t_{h45} \times BigM$$
$$\forall t \in T$$

(5.21)

$$P^t_{h54p} \leq u^t_{h54} \times BigM$$
$$\forall t \in T$$

(5.22)

$$P^t_{h45n} \leq u^t_{h54} \times BigM$$
$$\forall t \in T$$

(5.23)

The power from DES to homes should be less than the capacity of the line between the homes and the DES as modeled by Eq. (5.24). The load curtailment may be carried out on the loads as expressed by Eq. (5.25). The curtailed load is calculated by Eq. (5.26) and the load curtailment factor is defined by Eq. (5.27) [14].

$$\begin{cases} P^{t,h}_{home} \leq P^h_{rhome} \\ P^{t,h}_{home} \geq -P^h_{rhome} \end{cases} \quad \forall t \in T \quad \forall h \in H$$

(5.24)

$$P^{t,h}_{load} \times k^{t,h}_{lc} \geq P^{t,h}_{crload}$$
$$\forall t \in T$$
$$\forall h \in H$$

(5.25)

$$L^{t,h}_c = P^{t,h}_{load} \times \left(1 - k^{t,h}_{lc}\right)$$
$$\forall t \in T$$
$$\forall h \in H$$

(5.26)

$$k^{t,h}_{lc} \leq 1$$
$$\forall t \in T$$
$$\forall h \in H$$

(5.27)

The CGS is equipped with the diesel generator and battery. The power of the CGS is achieved as Eq. (5.28) which is formed by battery charging–discharging powers plus the produced power by the diesel generator. The rated capacity of the diesel generator is defined by Eq. (5.29). In practice, some diesel generators are not able to run continuously for 24-hour because of some issues like fuel tank capacity, heating issues, and maintenance. As a result, the operating hours of the diesel generator in one day (24-hour) and the number of hours the diesel generator can run continuously are limited. A binary variable is defined for the diesel generator operation in Eq. (5.30). By using this binary variable, the operating hours of diesel generator in 24-hour are limited as modeled in Eq. (5.31). The continuous operating hours of the diesel generator are modeled by Eq. (5.32).

$$P_{cgs}^t = P_{cb}^t - P_{db}^t - P_{dg}^t$$
$$\forall t \in T$$
(5.28)

$$P_{dg}^t \le P_{rdg}$$
$$\forall t \in T$$
(5.29)

$$P_{dg}^t \le u_{dg}^t \times BigM$$
$$\forall t \in T$$
(5.30)

$$\sum_{t=1}^{T} u_{dg}^t \le DG_{dot}$$
$$\forall t \in T$$
(5.31)

$$\left[u_{dg}^t + u_{dg}^{t-1} + u_{dg}^{t-2} + \cdots + u_{dg}^{t-k} \right] \le k$$
$$\forall t \in T, t \ne [1, 2, \cdots, k]$$
(5.32)

The battery charging–discharging powers in CGS are modeled by Eqs. (5.33)–(5.35). The binary variables are defined for charging–discharging powers in Eq. (5.33). By using these binary variables, the charging–discharging powers are modeled by Eqs. (5.34) and (5.35). The stored energy in the battery at each hour is calculated by Eq. (5.36). The initial energy inside the battery is given by Eq. (5.37). The rated capacity of the battery is defined in Eq. (5.38) and its efficiency is modeled by Eq. (5.39) [15].

$$u_{cb}^t + u_{db}^t \le 1$$
$$\forall t \in T$$
(5.33)

$$P_{cb}^t \le u_{cb}^t \times P_{rb}$$
$$\forall t \in T$$
(5.34)

$$P_{db}^t \le u_{db}^t \times P_{rb}$$
$$\forall t \in T$$
(5.35)

$$E_b^t = E_b^{t-1} + P_{cb}^t - \frac{P_{db}^t}{\eta_b}$$
$$\forall t \in T, t \ne [1]$$
(5.36)

$$E_b^t = E_b^0 + P_{cb}^t - \frac{P_{db}^t}{\eta_b}$$
$$\forall t = [1]$$
(5.37)

$$E_b^t \le E_{rb}$$
$$\forall t \in T$$
(5.38)

$$\eta_b = \frac{\displaystyle\sum_{t=1}^{T} P_{db}^t}{\displaystyle\sum_{t=1}^{T} P_{cb}^t}$$
(5.39)

5.4 Data of district energy system

The introduced DES is simulated by using typical data which are presented in this section. The parameters related to the line capacity, battery, and diesel generator can be found in Table 5.1. The diesel generator is allowed to operate 12 hours a day. As well, it is permitted to operate continuously for 3 hours. After 3 hours, the diesel generator needs to be switched off for one hour and it can afterward continue operation.

Table 5.2 lists the rated power of loads in each home as well as the rated capacity of generating system in each building. It is seen that homes 1 and 3 are supplied by the solar generating systems but home 5 is equipped with the wind generating system. The other homes only consume energy and do not possess energy resources. Table 5.3 presents the wind and solar power profiles in one typical day. The electricity price at each hour of the day is also presented in Table 5.3 [16].

Table 5.4 summarizes the 24-hour pattern of load power in each home. It is seen that different loading patterns are modeled for homes. The pattern of energy consumption by homes reveals that homes 1 and

TABLE 5.1 Parameters of the district energy system.

Parameter	Level
Rated power of line between DES and grid	100 kW
Rated power of diesel generator	25 kW
Rated power of battery	100 kW
Rated capacity of battery	200 kWh
Initial energy inside battery	0 kWh
Fuel cost for diesel generator operation	0.25 $/kWh
Penalty cost for load curtailment	1 $/kWh

TABLE 5.2 Rated power of loads and generating systems in each home.

	Rated power of load	Rated power of generating system
Home 1	10 kW	5 kW solar generating system
Home 2	15 kW	—
Home 3	25 kW	10 kW solar generating system
Home 4	10 kW	—
Home 5	20 kW	10 kW wind generating system
Home 6	20 kW	—

TABLE 5.3 Wind–solar powers and electricity price.

Hour	Wind power (%)	Solar power (%)	Electricity price ($/kWh)
1	6	0	0.10
2	56	0	0.10
3	82	0	0.10
4	52	0	0.10
5	44	0	0.10
6	70	0	0.10
7	76	5	0.10
8	82	10	0.10
9	84	15	0.15
10	84	25	0.15
11	100	40	0.15
12	100	60	0.15
13	78	80	0.20
14	64	100	0.20
15	100	90	0.15
16	92	80	0.15
17	84	70	0.20
18	80	50	0.20
19	78	30	0.20
20	32	10	0.20
21	4	5	0.20
22	8	0	0.20
23	10	0	0.15
24	0	0	0.15

4 are residential buildings while home 3 operates like commercial and business buildings in which the energy is consumed during business hours. Homes 5 consumes most of its energy during night hours. Homes 2 and 6 approximately consume a constant power. As a result, various loading patterns are modeled in the district. The critical loads of each home are presented in Table 5.5. The DES must supply these critical loads under any situation and event.

TABLE 5.4   Load power in each home.

Hour	Home 1 (%)	Home 2 (%)	Home 3 (%)	Home 4 (%)	Home 5 (%)	Home 6 (%)
1	15	30	5	10	70	90
2	15	30	5	10	90	90
3	15	30	5	10	90	85
4	10	30	5	5	80	80
5	10	30	5	5	60	90
6	19	30	5	15	50	85
7	20	30	5	15	70	90
8	20	30	90	20	90	95
9	30	50	100	20	100	100
10	40	50	95	30	80	100
11	60	50	100	40	70	90
12	70	70	90	50	60	95
13	90	70	100	70	40	85
14	90	70	95	70	20	80
15	80	70	100	60	15	90
16	60	100	100	50	15	95
17	50	100	95	50	10	100
18	70	90	90	80	10	85
19	90	90	20	100	15	100
20	100	80	20	100	20	90
21	80	80	5	90	25	90
22	70	60	5	60	30	95
23	60	60	5	30	40	90
24	30	60	5	15	60	85

5.5 Discussions on the test system

The developed DES is simulated and studied here. The daily costs are optimized as listed in Table 5.6. In this case, since there is not any outage or event, the DES does not utilize the diesel generator, the p2p operation, and the load curtailment. All the required energy is supplied by the external grid with the assistance of the central battery energy storage system. The daily energy cost of DES is 152.477 $/day.

Table 5.7 presents the exchanged power between the grid and the DES in various corridors. The received power from the external grid (i.e., P_{12}) is

TABLE 5.5 Critical load power in each home.

Hour	Home 1 (%)	Home 2 (%)	Home 3 (%)	Home 4 (%)	Home 5 (%)	Home 6 (%)
1	0	15	0	0	50	40
2	0	15	0	0	50	40
3	0	15	0	0	50	40
4	0	15	0	0	50	40
5	0	15	0	0	50	40
6	0	15	0	0	50	40
7	0	15	0	0	50	40
8	0	15	40	0	50	40
9	0	15	40	0	50	40
10	20	15	40	10	50	40
11	30	15	40	15	50	40
12	30	30	40	20	0	40
13	40	30	40	30	0	40
14	40	30	40	30	0	40
15	45	30	40	25	0	40
16	40	50	40	25	0	40
17	20	50	40	25	0	40
18	20	50	40	30	0	40
19	45	50	0	35	0	40
20	45	50	0	35	0	40
21	40	15	0	35	0	40
22	0	15	0	35	0	40
23	0	15	0	0	0	40
24	0	15	0	0	0	40

TABLE 5.6 Daily costs of district energy system.

Cost	Value
Purchasing power from grid ($/day)	152.477
Diesel generator operation ($/day)	0
Load curtailment penalty ($/day)	0
Energy losses in peer-to-peer exchange ($/day)	0
Total energy cost ($/day)	152.477

TABLE 5.7 Exchanged power between grid and district energy system.

Hour	P_{12} (kW)	P_{23} (kW)	P_{34} (kW)	P_{45} (kW)
1	100	97.5	79.6	60.35
2	38.65	36.15	19.25	0
3	35.05	32.55	18.25	0
4	34.05	32.55	17.25	0
5	100	98.5	86.4	67.15
6	29.15	25.75	18.25	0
7	100	96.75	85.85	67.1
8	63.7	60.2	45.9	5.4
9	66.85	62.6	43.5	0
10	62.1	56.35	41.25	0
11	58.5	50.5	39	0
12	57	48	35.5	0
13	−43.3	−55.3	−66	−100
14	47.65	36.65	28.55	−1.2
15	100	90.5	87	53
16	100	93	84.2	48.2
17	−48.15	−54.65	−63.25	−100
18	54.5	42	34.5	0
19	0.3	−17.2	−25.9	−47.9
20	54.3	34.8	22	0
21	100	83.25	66.65	47.9
22	−52.55	−65.55	−79.75	−100
23	44.25	35.25	19.25	0
24	43.75	39.25	18.25	0

positive in most of the hours which shows that the loads in the DES are supplied by the grid. As well, at some hours such as 13, 17, and 22 the DES sends power to the external grid. The CGS produces power at hours 13, 14, 17, 19, and 22. This produced power is consumed by homes in the DES and its surplus is injected into the external grid. For instance, the produced power by CGS at hours 14 and 19 is consumed by DES and the traded power with the external grid is positive which shows receiving power from the upstream grid. In such hours, the energy of DES is supplied from both directions of the grid and the CGS. Some loads of DES are supplied by the grid and some other loads are supported by the CGS. At hour 19, homes 2,

3, 5, and 6 are supplied by the CGS, and homes 1 and 4 are supplied by both the grid and the CGS. At hour 22, the produced power by CGS supplies all the homes, and the surplus of power is injected into the grid.

The flowed powers in the DES sectors at different hours are shown in Figs. 5.3 to 5.5. Fig. 5.3 demonstrates that the grid is supplying all homes as well as charges the battery in CGS. The p2p energy exchange is zero. Fig. 5.4 shows a condition in which the CGS is supplying all

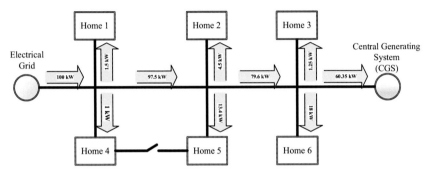

FIGURE 5.3　Flowed powers in district energy system at hour 1.

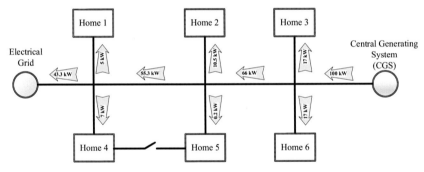

FIGURE 5.4　Flowed powers in district energy system at hour 13.

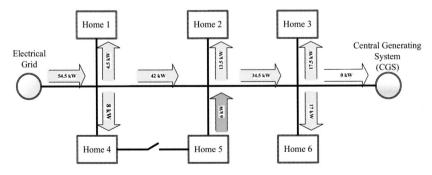

FIGURE 5.5　Flowed powers in DES at hour 18.

homes as well as sends power to the grid. In this condition, the DES sells energy to the grid. The p2p energy exchange is still zero. In Fig. 5.5, the grid is supplying all the homes excluding home 5. This home is integrated with a wind generating system and the surplus of wind energy is sent to the DES at hour 18. As a result, all homes consume energy but home 5 acts as a prosumer and generates electricity. The exchanged power with CGS is zero. In other words, the CGS neither generates electricity nor consumes energy.

Table 5.8 indicates the consumed-generated power by all homes in 24 hours. Homes 1 and 3 are supplied with solar panels and home 3

TABLE 5.8   Consumed-generated power by all homes.

Hour	Home 1 (kW)	Home 2 (kW)	Home 3 (kW)	Home 4 (kW)	Home 5 (kW)	Home 6 (kW)
1	1.5	4.5	1.25	1	13.4	18
2	1.5	4.5	1.25	1	12.4	18
3	1.5	4.5	1.25	1	9.8	17
4	1	4.5	1.25	0.5	10.8	16
5	1	4.5	1.25	0.5	7.6	18
6	1.9	4.5	1.25	1.5	3	17
7	1.75	4.5	0.75	1.5	6.4	18
8	1.5	4.5	21.5	2	9.8	19
9	2.25	7.5	23.5	2	11.6	20
10	2.75	7.5	21.25	3	7.6	20
11	4	7.5	21	4	4	18
12	4	10.5	16.5	5	2	19
13	5	10.5	17	7	0.2	17
14	4	10.5	13.75	7	−2.4	16
15	3.5	10.5	16	6	−7	18
16	2	15	17	5	−6.2	19
17	1.5	15	16.75	5	−6.4	20
18	4.5	13.5	17.5	8	−6	17
19	7.5	13.5	2	10	−4.8	20
20	9.5	12	4	10	0.8	18
21	7.75	12	0.75	9	4.6	18
22	7	9	1.25	6	5.2	19
23	6	9	1.25	3	7	18
24	3	9	1.25	1.5	12	17

is equipped with a wind turbine. These three prosumers can produce power at some hours. The other homes consume energy at all hours. The produced power by homes 1, 3, and 5 is used to feed their own loads, and the surplus of this energy is sent to the DES. If the produced power by the local resources is less than the load power, the transmitted power to the grid will be zero. This point is seen for homes 1 and 3. However, home 5 has a surplus of energy in hours 14−19 and the traded power with DES becomes negative at those hours.

The status of powers in home 1 is presented in Fig. 5.6 and the related numerical values of powers are listed in Table 5.9. It is clear that the produced solar power in this home is less than the load power at all hours. As a result, there is not any surplus of energy for trading with the grid and the received power from the grid is positive at all hours. Similar results are presented for home 3 in Fig. 5.7 and Table 5.10. In this home the produced power by the solar system is not sufficient to supply the loads and the home has to receive power from the grid. The home is not able to send power to the grid.

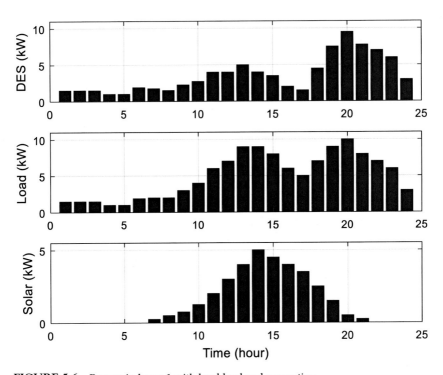

FIGURE 5.6 Powers in home 1 with local load and generation.

TABLE 5.9 Details of powers in home 1.

Hour	Power from DES (kW)	Load (kW)	Solar power (kW)
1	1.5	1.5	0
2	1.5	1.5	0
3	1.5	1.5	0
4	1	1	0
5	1	1	0
6	1.9	1.9	0
7	1.75	2	0.25
8	1.5	2	0.5
9	2.25	3	0.75
10	2.75	4	1.25
11	4	6	2
12	4	7	3
13	5	9	4
14	4	9	5
15	3.5	8	4.5
16	2	6	4
17	1.5	5	3.5
18	4.5	7	2.5
19	7.5	9	1.5
20	9.5	10	0.5
21	7.75	8	0.25
22	7	7	0
23	6	6	0
24	3	3	0

The status of powers in home 5 is presented in Fig. 5.8 and Table 5.11. In this home the produced power by the wind generating system supplies the loads and its surplus in hours 13–19 is traded with the grid. It is seen that the traded power with DES becomes negative at those mentioned hours.

The CGS is equipped with the battery and diesel generator. The diesel generator is not utilized by the DES but the battery is used properly. The battery operations including charging–discharging power and

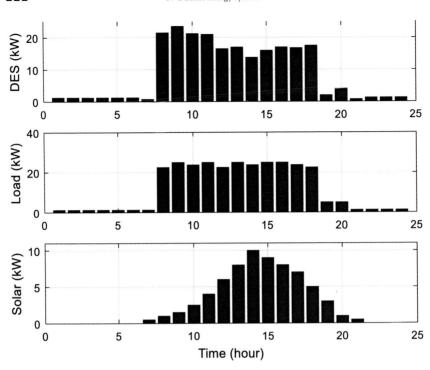

FIGURE 5.7 Powers in home 3 with local load and generation.

TABLE 5.10 Details of powers in home 3.

Hour	Power from DES (kW)	Load (kW)	Solar power (kW)
1	1.25	1.25	0
2	1.25	1.25	0
3	1.25	1.25	0
4	1.25	1.25	0
5	1.25	1.25	0
6	1.25	1.25	0
7	0.75	1.25	0.5
8	21.5	22.5	1
9	23.5	25	1.5
10	21.25	23.75	2.5
11	21	25	4
12	16.5	22.5	6

(Continued)

Energy Management in Homes and Residential Microgrids

TABLE 5.10 (Continued)

Hour	Power from DES (kW)	Load (kW)	Solar power (kW)
13	17	25	8
14	13.75	23.75	10
15	16	25	9
16	17	25	8
17	16.75	23.75	7
18	17.5	22.5	5
19	2	5	3
20	4	5	1
21	0.75	1.25	0.5
22	1.25	1.25	0
23	1.25	1.25	0
24	1.25	1.25	0

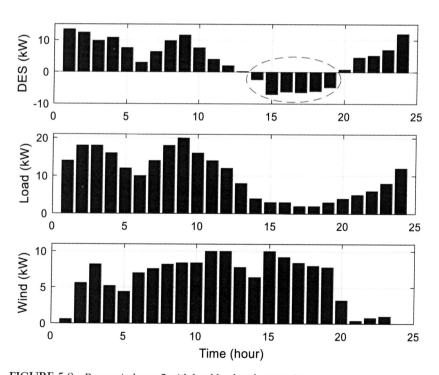

FIGURE 5.8 Powers in home 5 with local load and generation.

TABLE 5.11 Details of powers in home 5.

Hour	Power from DES (kW)	Load (kW)	Wind power (kW)
1	13.4	14	0.6
2	12.4	18	5.6
3	9.8	18	8.2
4	10.8	16	5.2
5	7.6	12	4.4
6	3	10	7
7	6.4	14	7.6
8	9.8	18	8.2
9	11.6	20	8.4
10	7.6	16	8.4
11	4	14	10
12	2	12	10
13	0.2	8	7.8
14	−2.4	4	6.4
15	−7	3	10
16	−6.2	3	9.2
17	−6.4	2	8.4
18	−6	2	8
19	−4.8	3	7.8
20	0.8	4	3.2
21	4.6	5	0.4
22	5.2	6	0.8
23	7	8	1
24	12	12	0

energy are demonstrated in Fig. 5.9. It is seen that the battery is operated for many hours and shifts energy over 24-hour of the day. The battery is fully charged in hours 7 to 12 and such energy is discharged and used at the next hours like hour 13. The battery charging–discharging scheduling assists the DES to minimize the daily energy cost. However, it can also help the DES to supply the loads when the DES is faced with outages and events. This point is discussed in the next sections.

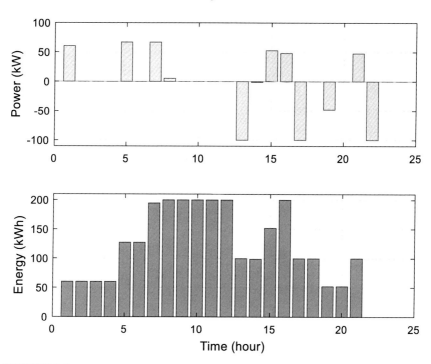

FIGURE 5.9 Charging–discharging power of battery together with its energy.

5.6 Outage of line 23

As already mentioned, the DES is supplied from two directions of the grid and the CGS. The CGS utilizes the battery and the diesel generator for supplying the DES. The DES may be supplied by either the grid or the CGS or even both of them at the same time. The CGS not only helps the DES to reduce the daily energy cost but also is able to assist the DES to cope with outages and events. In order to demonstrate the ability of CGS to deal with events, line 23 is subjected to an outage in hours 11–15 and the DES is simulated under such a situation. The daily energy cost of DES is presented in Table 5.12. It is clear that the DES uses p2p operation to deal with the outage of the line. The diesel generator and load curtailment are not utilized and the DES deals with the outage by appropriate utilization of battery. As a result, the daily energy cost increases just by a small value.

Following the aforementioned outage, the DES is sectionalized into two separate parts. The left part is supplied by the grid and the right part is fed by the CGS. The connection between those two parts is made by p2p operation. Table 5.13 presents the flowed power in different sectors of DES following the outage of line 23 in hours 11–15. Under this outage, the DES is supplied from both directions. The loads on node 1 (homes 1 and 4) are

TABLE 5.12   Daily costs of district energy system following the outage of line 23.

Cost	Value
Purchasing power from grid ($/day)	161.177
Diesel generator operation ($/day)	0.000
Load curtailment penalty ($/day)	0.000
Energy losses in peer-to-peer exchange ($/day)	0.148
Total energy cost ($/day)	161.325

TABLE 5.13   Exchanged power between grid and DES following outage of line 23.

Hour	P_{12} (kW)	P_{23} (kW)	P_{34} (kW)	P_{45} (kW)
1	100	97.5	79.6	60.35
2	−1.35	−3.85	−20.75	−40
3	35.05	32.55	18.25	0
4	34.05	32.55	17.25	0
5	32.85	31.35	19.25	0
6	100	96.6	89.1	70.85
7	100	96.75	85.85	67.1
8	100	96.5	82.2	41.7
9	66.85	62.6	43.5	0
10	62.1	56.35	41.25	0
11	14	0	−5.5	−44.5
12	14	0	−7.5	−43
13	15	0	−7.7	−41.7
14	11.55	0	−7.55	−37.3
15	13.5	0	0.5	−33.5
16	100	93	84.2	48.2
17	3.65	−2.85	−11.45	−48.2
18	54.5	42	34.5	0
19	48.2	30.7	22	0
20	54.3	34.8	22	0
21	52.1	35.35	18.75	0
22	47.45	34.45	20.25	0
23	44.25	35.25	19.25	0
24	43.75	39.25	18.25	0

supplied by the grid and the rest of the homes are supplied by CGS. The flowed power through line 23 (P_{23}) is zero. In the rest of the hours when the DES is not facing an outage, the CGS assists the DES to reduce the operating cost by proper charging scheduling of the battery.

The p2p operation from home 4 to home 5 is shown in Fig. 5.10. Throughout the event, the power is transferred from home 4 to home 5 to deal with the outage of the line. The injected power to the home may be consumed by its loads or can be injected into the DES. The powers of homes 4 and 5 are listed in Table 5.14. The results verify that the injected power from home 4 to home 5 is afterward sent to the DES. In other words, home 5 receives power from home 4 and trades it with DES in order to handle the outage of line 23.

Fig. 5.11 depicts the flowed powers in various sectors of DES at hour 13. It is seen that the DES is supplied from two directions by grid and CGS. Homes 1 and 4 are supplied by the grid and homes 2, 3, 5, and 6 are supplied by the CGS. The p2p energy exchange between home 4 and home 5 is utilized by DES and 3 kW power is sent from home 4 to home 5. A small part of this power is consumed by home 5 (i.e., 0.2 kW) and 2.8 kW of power is injected to DES for supplying home 2.

Fig. 5.12 demonstrates the power and energy of the battery following the outage of line 23 at hours 11−15. During the outage hours, the battery works in the discharging state and produces energy for DES. In the

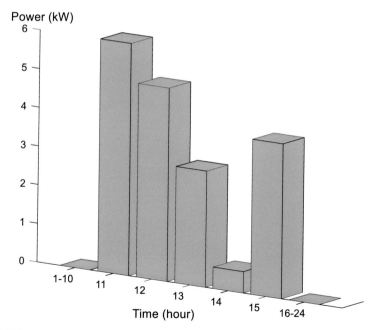

FIGURE 5.10   Power exchange from home 4 to home 5 following outage of line 23.

TABLE 5.14 Power of homes 4 and 5 following outage of line 23.

Hour	Home 4 (kW)	Home 5 (kW)
1	1	13.4
2	1	12.4
3	1	9.8
4	0.5	10.8
5	0.5	7.6
6	1.5	3
7	1.5	6.4
8	2	9.8
9	2	11.6
10	3	7.6
11	10	−2
12	10	−3
13	10	−2.8
14	7.55	−2.95
15	10	−11
16	5	−6.2
17	5	−6.4
18	8	−6
19	10	−4.8
20	10	0.8
21	9	4.6
22	6	5.2
23	3	7
24	1.5	12

rest of the hours, the battery operation is scheduled to decrease the operating cost in the DES. For instance, the battery is charged at hour 15 and discharged at hour 16 to decrease the energy cost.

5.7 Outage of line 34

The outage of line 34 in hours 16–19 is modeled as another incident. Following the outage of line 34, the DES is supplied from both directions

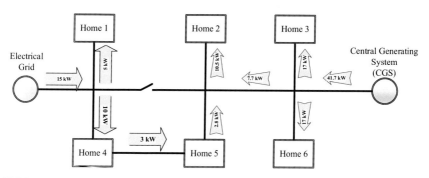

FIGURE 5.11   Flowed powers in DES at hour 13 following outage of line 23.

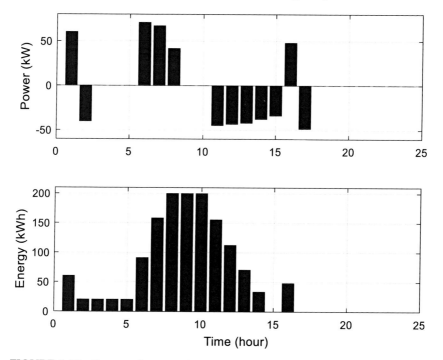

FIGURE 5.12   Power and energy of battery following outage of line 23.

by two resources including the grid and the CGS. The daily costs are listed in Table 5.15. The diesel generator and load curtailment are not utilized and the DES deals with the outage by proper utilization of battery. As a result, the daily energy does not show a significant increase compared to the first case. The p2p operation is not used by DES as well.

The flowed powers in various sectors of DES are presented in Table 5.16. During the outage of line 34, the flowed power through this line is zero. Homes 3 and 6 are connected to CGS and the other homes

TABLE 5.15 Daily costs of district energy system following outage of line 34.

Cost	Value
Purchasing power from grid ($/day)	156.68
Diesel generator operation ($/day)	0
Load curtailment penalty ($/day)	0
Energy losses in peer-to-peer exchange ($/day)	0
Total energy cost ($/day)	156.68

TABLE 5.16 Exchanged power between grid and district energy system following outage of line 34.

Hour	P_{12} (kW)	P_{23} (kW)	P_{34} (kW)	P_{45} (kW)
1	100	97.5	79.6	60.35
2	100	97.5	80.6	61.35
3	35.05	32.55	18.25	0
4	100	98.5	83.2	65.95
5	32.85	31.35	19.25	0
6	29.15	25.75	18.25	0
7	45.25	42	31.1	12.35
8	58.3	54.8	40.5	0
9	66.85	62.6	43.5	0
10	62.1	56.35	41.25	0
11	58.5	50.5	39	0
12	57	48	35.5	0
13	32.95	20.95	10.25	−23.75
14	−51.15	−62.15	−70.25	−100
15	100	90.5	87	53
16	15.8	8.8	0	−36
17	15.1	8.6	0	−36.75
18	20	7.5	0	−34.5
19	26.2	8.7	0	−22
20	54.3	34.8	22	0
21	52.1	35.35	18.75	0
22	47.45	34.45	20.25	0
23	44.25	35.25	19.25	0
24	43.75	39.25	18.25	0

are integrated to the grid. The p2p energy exchange is not used. The DES uses the battery to supply the loads in homes 3 and 6.

Fig. 5.13 shows the flowed powers in DES at hour 18. It is seen that the DES is divided into two separate sections. Homes 1, 2, 4, and 5 are supplied by the grid, and homes 3 and 6 are supplied by the CGS. The introduced farmwork properly deals with the outage of the line and converts the DES into two separate sections and supplies each part separately. The battery operation is depicted in Fig. 5.14. The operation of the battery shows the discharging process in hours 16–19 when the DES is facing an outage of line 34.

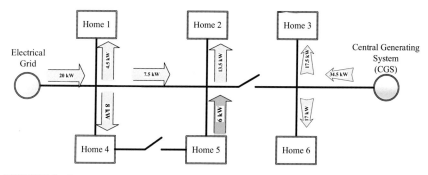

FIGURE 5.13   Flowed powers in district energy system at hour 18 following outage of line 34.

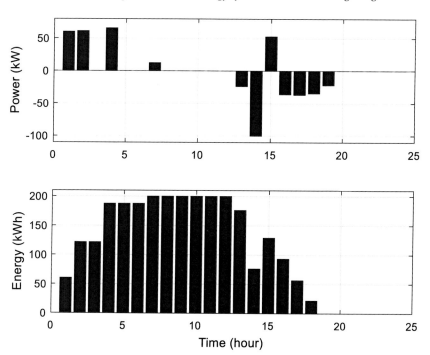

FIGURE 5.14   Power and energy of battery following outage of line 34.

5.8 Outage of line 45

During the outage of line 45 in hours 6–17, the DES is only connected to the external grid and CGS is disconnected from the DES. All the homes should be supplied by the grid. The daily energy cost is presented in Table 5.17 and it does not show considerable change compared to the previous cases because all the loads are supplied by the grid and the local options (diesel generator, p2p operation, and load curtailment) are not utilized by the DES.

Table 5.18 summarizes the exchanged power between the grid and the DES following an outage of line 45 and it is clear that P_{45} is zero for the duration of the outage, where, the CGS cannot participate in the energy management of DES and all loads are supplied by the grid. The operation of CGS during the outage is demonstrated in Fig. 5.15. The whole time of the outage, the battery has no operation and only keeps the stored energy for the next hours. The carried energy by the battery is consumed at hours 18–19.

TABLE 5.17 Daily costs of district energy system following outage of line 45.

Cost	Value
Purchasing power from grid ($/day)	157.53
Diesel generator operation ($/day)	0
Load curtailment penalty ($/day)	0
Energy losses in peer-to-peer exchange ($/day)	0
Total energy cost ($/day)	157.53

TABLE 5.18 Exchanged power between grid and DES following outage of line 45.

Hour	P_{12} (kW)	P_{23} (kW)	P_{34} (kW)	P_{45} (kW)
1	41.6	39.1	21.2	1.95
2	38.65	36.15	19.25	0
3	100	97.5	83.2	64.95
4	100	98.5	83.2	65.95
5	100	98.5	86.4	67.15
6	29.15	25.75	18.25	0
7	32.9	29.65	18.75	0
8	58.3	54.8	40.5	0
9	66.85	62.6	43.5	0
10	62.1	56.35	41.25	0

(Continued)

TABLE 5.18   (Continued)

Hour	P_{12} (kW)	P_{23} (kW)	P_{34} (kW)	P_{45} (kW)
11	58.5	50.5	39	0
12	57	48	35.5	0
13	56.7	44.7	34	0
14	48.85	37.85	29.75	0
15	47	37.5	34	0
16	51.8	44.8	36	0
17	51.85	45.35	36.75	0
18	−45.5	−58	−65.5	−100
19	−51.8	−69.3	−78	−100
20	54.3	34.8	22	0
21	52.1	35.35	18.75	0
22	47.45	34.45	20.25	0
23	44.25	35.25	19.25	0
24	43.75	39.25	18.25	0

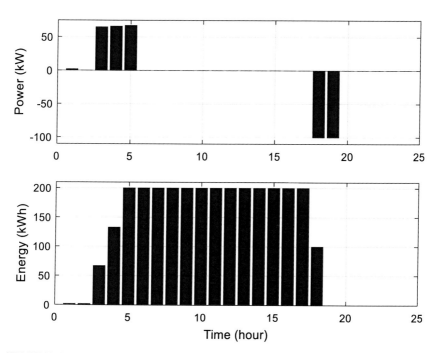

FIGURE 5.15   Power and energy of battery following outage of line 45.

5.9 Outage of line 12 in hours 14−24

The outage of line 12 in hours 14−24 is modeled as an event. The outage of line 12 leads to the off-grid operation of the DES. In such a situation, the DES is just connected to the CGS and all the required energy for homes should be supplied by the CGS. Table 5.19 presents the daily costs and it is seen that the DES utilizes all available resources to deal with such a situation. The purchased power from the grid reduces, but the costs of diesel generator and load curtailment increase.

Table 5.20 lists the flowed powers in the DES following an outage of line 12. During the outage of line 12, the exchanged power with the grid is zero and the DES operates off-grid. In hours 14−24, all the homes are supplied by CGS resources including the battery and the diesel generator. As well, the load curtailment option is carried out and some loads are curtailed.

TABLE 5.19 Daily costs of district energy system following outage of line 12.

Cost	Value
Purchasing power from grid ($/day)	98.067
Diesel generator operation ($/day)	56.250
Load curtailment penalty ($/day)	119.050
Energy losses in peer-to-peer exchange ($/day)	0.000
Total energy cost ($/day)	273.367

TABLE 5.20 Exchanged power between grid and district energy system following outage of line 12.

Hour	P_{12} (kW)	P_{23} (kW)	P_{34} (kW)	P_{45} (kW)
1	100	97.5	79.6	60.35
2	38.65	36.15	19.25	0
3	−4.95	−7.45	−21.75	−40
4	34.05	32.55	17.25	0
5	32.85	31.35	19.25	0
6	100	96.6	89.1	70.85
7	100	96.75	85.85	67.1
8	100	96.5	82.2	41.7
9	66.85	62.6	43.5	0

(Continued)

TABLE 5.20   (Continued)

Hour	P_{12} (kW)	P_{23} (kW)	P_{34} (kW)	P_{45} (kW)
10	62.1	56.35	41.25	0
11	58.5	50.5	39	0
12	57	48	35.5	0
13	56.7	44.7	34	0
14	0	−2	−6	−35.75
15	0	−6	−6.5	−40.5
16	0	−4.5	−10.3	−46.3
17	0	−4	−10.6	−47.35
18	0	−7.5	−13	−47.5
19	0	−11	−10.7	−25
20	0	−13	−21.8	−43.8
21	0	−11.25	−22.85	−41.6
22	0	−10.5	−18.7	−38.95
23	0	−6	−14	−33.25
24	0	−2.75	−17	−25

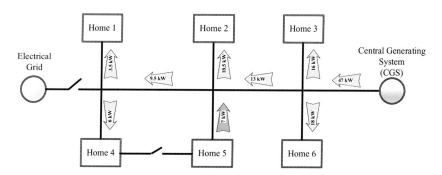

FIGURE 5.16   Off-grid operation of DES at hour 15.

Fig. 5.16 shows the off-grid operation of DES at hour 15. It is seen that the exchanged power with the external grid is zero. All homes are supplied by CGS. The diesel generator and battery in CGS operate to supply the required energy of loads.

Table 5.21 summarizes the load curtailment in homes and it is obvious that the load shedding is applied when the DES faces an outage on line 12. Homes 4 and 5 suffer the most load curtailment because of their

TABLE 5.21 Load curtailment in homes per kW.

Hour	Home 1 (kW)	Home 2 (kW)	Home 3 (kW)	Home 4 (kW)	Home 5 (kW)	Home 6 (kW)
1–13	0	0	0	0	0	0
14	5	0.1	0	4	4	0
15	0	0	0	3.5	3	0
16	0	0	0	2.5	3	0
17	0	0	0	2.5	2	0
18	0	0	0	5	2	0
19	0	6	5	6.5	3	2.7
20	0	0	0	6.5	4	0
21	0	0	0	5.5	5	0
22	0	0	0	2.5	6	0
23	0	0	0	3	8	0
24	1.75	6.75	1.25	0	0	9

higher load levels. As already stated, each home has two parts of loads including critical and noncritical loads. The critical loads must be supplied under all situations while the noncritical loads can be curtailed when necessary. The applied load curtailment only deals with noncritical loads as shown in Fig. 5.17, where, the supplied loads and the critical loads are shown. It is seen that the supplied loads in all hours are greater than or equal to the critical loads. That point approves supplying the critical loads at all hours of the day even during outage hours.

The resources in the CGS including the diesel generator and the battery are properly operated to handle such a situation. The diesel generator operation is depicted in Fig. 5.18 and the battery operation is displayed in Fig. 5.19. Once the DES becomes off-grid, both of the resources produce and inject power to the DES. The diesel generator operates at the maximum capacity to avoid load curtailment.

5.10 Off-grid operation

In this case, it is assumed that the DES is disconnected from the grid for 24-hour and has off-grid operation for the whole day. The initial energy of the battery in the CGS is 100 kWh. The daily energy cost is presented in Table 5.22. The grid cost is zero and the diesel generator is

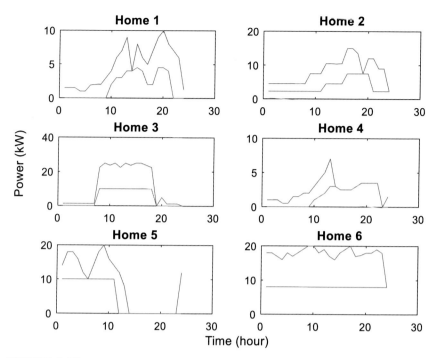

FIGURE 5.17   Supplied and critical loads under outage of line 12. *Red line*: critical load power; *Blue line*: supplied load power.

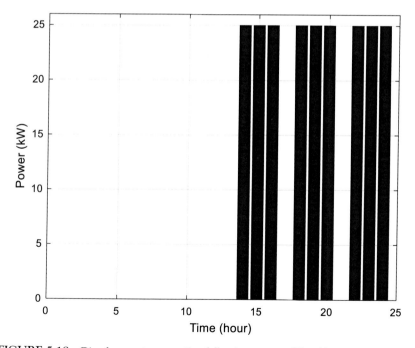

FIGURE 5.18   Diesel generator operation following outage of line 12.

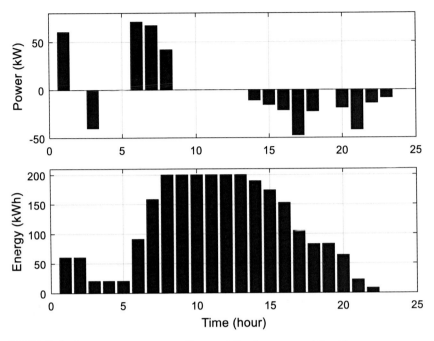

FIGURE 5.19   Power and energy of battery following outage of line 12.

TABLE 5.22   Daily costs of off-grid district energy system.

Cost	Value
Purchasing power from grid ($/day)	0.000
Diesel generator operation ($/day)	75.000
Load curtailment penalty ($/day)	745.800
Energy losses in peer-to-peer exchange ($/day)	0.000
Total energy cost ($/day)	820.800

operated significantly. The diesel generator and battery in the CGS attempt to supply all the loads, however, when the CGS is not able to deal with the loads, the load curtailment is applied.

Table 5.23 presents the flowed powers in the off-grid DES. The traded power with the external grid is zero at all hours. The CGS produces energy at all hours to handle loads of homes. The curtailed load in each home is shown in Table 5.24. It is obvious that the loads are curtailed when the CGS is not able to supply them.

TABLE 5.23 Exchanged power between external grid and off-grid district energy system.

Hour	P_{12} (kW)	P_{23} (kW)	P_{34} (kW)	P_{45} (kW)
1	0	0	−11.65	−19.65
2	0	0	−6.65	−14.65
3	0	0	−4.05	−12.05
4	0	−1	−10.3	−25
5	0	0	−7.85	−15.85
6	0	0	−5.25	−13.25
7	0	0.25	−4.4	−11.9
8	0	0.5	−3.55	−20.55
9	0	0.75	−3.1	−25
10	0	−3.75	−7.6	−25
11	0	−2.5	−4.75	−18.75
12	0	−5	0.5	−25
13	0	−3	0.3	−9.7
14	0	−2	−0.1	−8.1
15	0	−2.5	3	−6
16	0	−2.5	−0.8	−10.8
17	0	−1	−0.1	−11.1
18	0	−7.5	−9	−25
19	0	−9.65	−9.35	−14.35
20	0	−7.5	−11.8	−18.8
21	0	−11.25	−13.1	−25
22	0	−10.5	−11.95	−25
23	0	0	−1.25	−9.25
24	0	0	−2.25	−10.25

The collaborative operation between battery and diesel generator is also interesting to be evaluated. Table 5.25 presents the battery and diesel generator operation. As already modeled and discussed, the diesel generator is not able to operate for more than 3 consecutive hours. When the diesel generator needs to be switched off for a while, the battery takes over its duty and supplies the loads. As a result, the battery

TABLE 5.24 Load curtailment per kW in off-grid district energy system.

Hour	Home 1 (kW)	Home 2 (kW)	Home 3 (kW)	Home 4 (kW)	Home 5 (kW)	Home 6 (kW)
1	1.5	2.25	1.25	1	4	10
2	1.5	2.25	1.25	1	8	10
3	1.5	2.25	1.25	1	8	9
4	0	0	0	0.5	6	2.55
5	1	2.25	1.25	0.5	2	10
6	1.9	2.25	1.25	1.5	0	9
7	2	2.25	1.25	1.5	4	10
8	2	2.25	12.5	2	8	11
9	3	5.25	9.6	2	10	12
10	0	5.25	11.85	2	6	12
11	3	5.25	15	2.5	4	10
12	4	6	0	0	12	10
13	5	6	15	4	8	9
14	5	6	13.75	4	4	8
15	3.5	6	15	3.5	3	10
16	2	7.5	15	2.5	3	11
17	3	7.5	13.75	2.5	2	12
18	0	6	9.5	5	0	9
19	1.35	6	5	6.5	3	12
20	5.5	4.5	5	6.5	4	10
21	0	9.75	1.25	5.5	5	5.6
22	0	6.75	1.25	2.5	6	5.95
23	6	6.75	1.25	3	8	10
24	3	6.75	1.25	1.5	12	9

should be optimally charged for collaborative operation with the diesel generator. This point is seen in the results of the simulations. For instance, the battery is charged at hour 6 and discharged at hours 7–8 when the diesel generator is switched off.

Fig. 5.20 indicates the supplied loads and the critical loads in off-grid DES. The supplied loads in all hours are greater than or equal to

TABLE 5.25 Battery and diesel generator operation.

Hour	Diesel generator (kW)	Battery charge (kW)	Battery discharge (kW)	Battery energy (kWh)
1	0	0	19.65	80.35
2	25	10.35	0	90.7
3	25	12.95	0	103.65
4	25	0	0	103.65
5	0	0	15.85	87.8
6	25	11.75	0	99.55
7	0	0	11.9	87.65
8	0	0	20.55	67.1
9	25	0	0	67.1
10	25	0	0	67.1
11	0	0	18.75	48.35
12	25	0	0	48.35
13	0	0	9.7	38.65
14	0	0	8.1	30.55
15	0	0	6	24.55
16	25	14.2	0	38.75
17	0	0	11.1	27.65
18	25	0	0	27.65
19	25	10.65	0	38.3
20	0	0	18.8	19.5
21	25	0	0	19.5
22	25	0	0	19.5
23	0	0	9.25	10.25
24	0	0	10.25	0

the critical loads. It is verified that the DES is able to supply all the critical loads over 24-hour in the off-grid operation. In some states, the supplied load is exactly equal to the critical load which means that the DES is only able to supply the critical load, and the rest of the loads are curtailed.

5.11 Peer-to-peer energy exchange

In order to demonstrate the p2p energy exchange, home 5 is disconnected from DES in hours 6−19. Under such a condition, the only way to supply home 5 is the p2p energy exchange with home 4. In this case, the central battery in CGS is modeled with 100 kWh of initial energy. The daily energy costs are listed in Table 5.26 and the p2p operation is

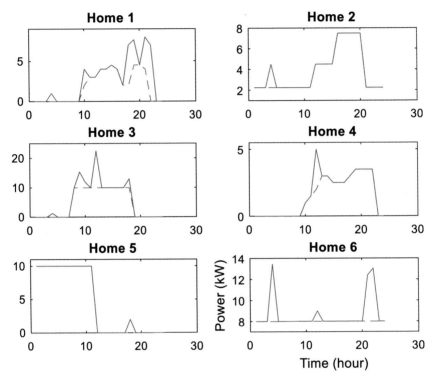

FIGURE 5.20 Supplied and critical loads under off-grid district energy system. *Red line*: critical load power; *Blue line*: supplied load power.

TABLE 5.26 Daily costs of district energy system following outage of home 5 in hours 6−19.

Cost	Value
Purchasing power from grid ($/day)	141.667
Diesel generator operation ($/day)	0.000
Load curtailment penalty ($/day)	6.000
Energy losses in peer-to-peer exchange ($/day)	0.541
Total energy cost ($/day)	148.209

used by DES. As well, some loads are curtailed and their power cannot be supplied wholly.

The p2p energy exchange between home 4 and home 5 is depicted in Fig. 5.21. The power is transferred from home 4 to home 5 in hours 6−13 and it is transferred back from home 5 to home 4 in hours 14−19. In hours 6−13, the energy of loads in home 5 is supplied through home 4. However, home 5 is equipped with wind turbines and the excess wind energy can be used by home 4 or DES. This point is seen in hours 14−19 when the direction of power is reversed and home 5 sends excess wind energy to home 4. The p2p energy exchange not only provides a parallel way to supply loads of islanded homes but also assists the homes in exchanging surplus energy with the adjacent homes. In such a situation, the excess energy in home 5 is not wasted and it is used by DES. If the p2p energy exchange is not used, the surplus of wind energy in home 5 should be wasted or it needs further facilities such as energy storage systems to be stored and used.

As shown by Table 5.27, The load curtailment is applied in home 5 at hours 8−10 when the wind energy and p2p energy exchange are not adequate to supply the loads. However, the critical loads are still supplied in home 5 as depicted by Fig. 5.22, where, the supplied load at each hour is greater than the critical load.

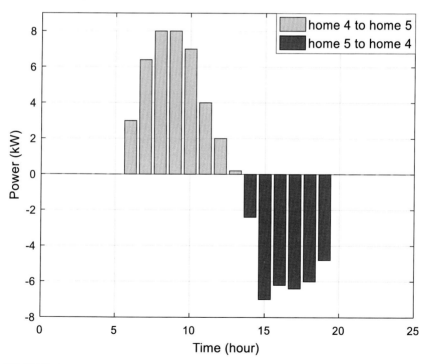

FIGURE 5.21   Peer-to-peer energy exchange between home 4 and home 5.

TABLE 5.27 Load curtailment per kW under outage of home 5 at hours 6–19.

Hour	Homes 1 to 4 (kW)	Home 5 (kW)	Home 6 (kW)
1–7	0	0	0
8	0	1.800	0
9	0	3.600	0
10	0	0.600	0
11–24	0	0	0

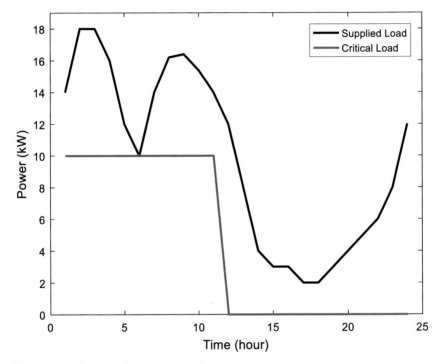

FIGURE 5.22   Critical load and supplied load in home 5.

5.12 24-hour outage of home 5

When home 5 is disconnected from the DES for 24 hours, the only way to supply this home is the p2p energy exchange with home 4. If the taken power from home 4 is not enough to supply the loads, the load curtailment option may be employed by DES. The daily costs following the outage of home 5 for 24-hour are listed in Table 5.28. It is obvious that p2p energy exchange and load curtailment are utilized by

TABLE 5.28 Daily costs following 24-hour outage of home 5.

Cost	Value
Purchasing power from grid ($/day)	149.032
Diesel generator operation ($/day)	0.000
Load curtailment penalty ($/day)	25.000
Energy losses in peer-to-peer exchange ($/day)	0.930
Total energy cost ($/day)	174.963

DES. The energy management system supplies loads of home 5 via home 4 as much as possible and curtails the rest of the loads. It should be mentioned that the load curtailment is applied on the noncritical loads and the DES is obligated to supply all the critical loads.

Table 5.29 presents the load curtailment in DES following 24-hour outage of home 5. It is seen that most of the load curtailment is applied on home 5. A small load curtailment is applied on home 4 at hour 1. At this hour, the exchanged power from home 4 to home 5 is not enough to supply the critical loads of home 5. As a result, the DES curtails 0.4 kW of noncritical loads in home 4 and devotes the achieved energy to supply the critical loads in home 5.

Fig. 5.23 demonstrates the p2p power from home 4 to home 5 following 24-hour outage of home 5. The power is sent from home 4 to home 5 in most of the hours but hours 14−19. At these hours, the wind generating system in home 5 supplies the loads, and the excess of wind energy is sent to home 4. The p2p energy exchange not only serves the loads on home 5 but also harvests the surplus of energy in home 5 and avoids wasting energy. Without the p2p operation, neither the loads in home 5 are supplied nor the surplus of wind energy can be used.

5.13 Uncertainty in parameters

In order to model the uncertainty in the parameters, three parameters of load power, wind energy, and solar energy are assumed as uncertain parameters. Some relationships of the deterministic model are re-formulated for achieving a stochastic model. The cost of purchased electricity from the external grid is re-formulated as Eq. (5.40). The load curtailment cost is defined as Eq. (5.41).

$$Z_{grid} = \sum_{s=1}^{S} \left\{ \left\{ \sum_{t=1}^{T} \left(P_{grid}^t \times E_{price}^t \right) \right\} \times O_{pr}^s \right\} \tag{5.40}$$

TABLE 5.29 Load curtailment following 24-hour outage of home 5.

Hour	Home 4 (kW)	Home 5 (kW)
1	0.4	4
2	0	3.4
3	0	0.8
4	0	1.3
5	0	0
6	0	0
7	0	0
8	0	1.8
9	0	3.6
10	0	0.6
11	0	0
12	0	0
13	0	0
14	0	0
15	0	0
16	0	0
17	0	0
18	0	0
19	0	0
20	0	0.8
21	0	3.6
22	0	1.2
23	0	0
24	0	3.5

$$Z_{lc} = \sum_{s=1}^{S} \left\{ \left\{ \left(\sum_{h=1}^{H} \left[\sum_{t=1}^{T} \left(L_c^{t,h} \right) \right] \right) \times C_{cost} \right\} \times O_{pr}^{s} \right\} \qquad (5.41)$$

The power from the grid toward the DES is expressed by Eq. (5.42) and it is seen that the power is associated with uncertainty index "s." The flowed power through each sector of DES is calculated by

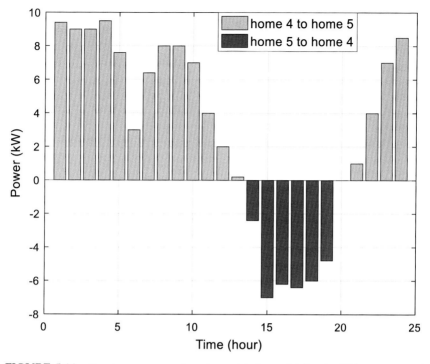

FIGURE 5.23   Peer-to-peer power from home 4 to home 5 following 24-hour outage of home 5.

Eq. (5.43). The power between CGS and DES is addressed by Eq. (5.44) and this power is not associated with uncertainty. In other words, the CGS provides a unique operating pattern under all scenarios of performance. The CGS is equipped with the diesel generator and battery. These two resources have a unique operating pattern that is constant under all uncertainties in the parameters. The traded power between each home and DES is presented by Eq. (5.45). Except for these relationships, the other formulations are similar to the deterministic model already introduced.

$$P_{grid}^{t,s} = P_{12}^{t,s}$$
$$\forall t \in T$$
$$\forall s \in S$$
(5.42)

$$P_i^{t,s} = P_{i+1}^{t,s} + \sum_{h=1}^{H} P_{home}^{t,s,h}$$
$$\forall t \in T$$
$$\forall s \in S$$
(5.43)

$$P_{45}^t = P_{cgs}^t$$
$$\forall t \in T \tag{5.44}$$

$$P_{home}^{t,h,s} = P_{load}^{t,h,s} \times K_{lc}^{t,h,s} - P_{solar}^{t,h,s} - P_{wind}^{t,h,s}$$
$$\forall t \in T$$
$$\forall s \in S \tag{5.45}$$
$$\forall h \in H$$

Seven scenarios of performance are defined to model the uncertainty in the parameters. The solar and wind energy scenarios are taken from Chapter 4. Loads of some homes under all scenarios of performance are depicted in Fig. 5.24. In these plots, the black solid line demonstrates the main scenario of performance which shows the deterministic operating condition. The alterations of loads in each scenario are presented relative to the main scenario of performance. In order to provide more details, the numerical loading data in home 6 under all scenarios of performance are presented in Table 5.30. Scenario 1 is assumed as the main scenario of performance which shows the deterministic model. The

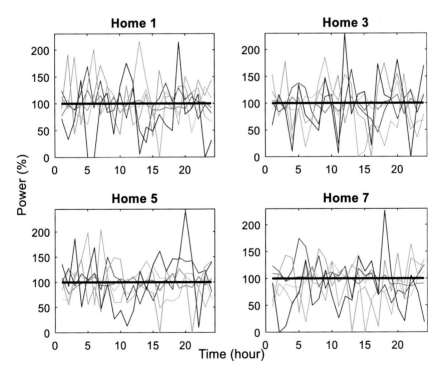

FIGURE 5.24   Alterations of load in each scenario relative to main scenario of performance Black solid line: main scenario of performance.

TABLE 5.30 Load of home 6 in percent under all scenarios of performance.

Hour	Scenario number						
	1	2	3	4	5	6	7
1	100	109	91	124	57	48	92
2	100	108	67	113	112	91	0
3	100	95	90	95	143	77	12
4	100	100	85	116	138	52	55
5	100	103	84	175	36	78	74
6	100	102	94	159	129	19	135
7	100	120	123	98	93	92	50
8	100	84	100	142	155	90	62
9	100	109	108	81	137	58	115
10	100	112	132	111	86	132	66
11	100	94	114	104	111	48	26
12	100	105	103	123	120	88	67
13	100	99	122	64	2	126	59
14	100	90	145	119	75	111	141
15	100	101	90	105	84	0	113
16	100	117	117	110	54	61	133
17	100	96	84	69	11	80	72
18	100	106	75	88	134	72	227
19	100	89	86	54	75	32	107
20	100	91	99	78	147	98	45
21	100	88	61	71	62	92	21
22	100	86	68	90	138	164	106
23	100	91	99	105	128	106	76
24	100	91	112	125	122	135	19

other scenarios form the uncertainty and their alterations are presented relative to the main scenario. Fig. 5.25 shows the probability of occurrence for each scenario of performance. In the stochastic model, the load power changes and it would exceed the line capacity. As a result, the capacities of connecting lines between homes and DES are increased by

100% to deal with such uncertainty. The rest of the data are taken into account similar to the deterministic model.

The developed stochastic model is simulated on the test DES and the daily costs are listed in Table 5.31. It is seen that the diesel generator and p2p energy exchange are not utilized by the DES. The powers of all homes are supplied by the grid and some loads are faced with curtailment. The daily cost is 162.6 $/day which shows about 6% growth compared to the determinist model.

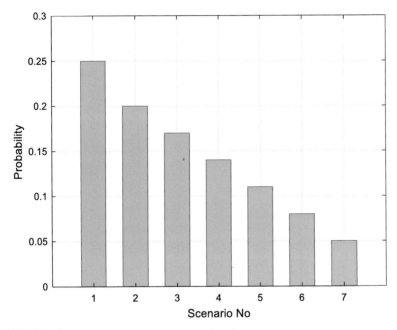

FIGURE 5.25   Probability of each scenario of performance.

TABLE 5.31   Daily costs of DES under parametric uncertainty.

Cost	Value
Purchasing power from grid ($/day)	162.433
Diesel generator operation ($/day)	0.000
Load curtailment penalty ($/day)	0.172
Energy losses in peer-to-peer exchange ($/day)	0.000
Total energy cost ($/day)	162.605

The DES comprises four sections and the flowed power in each sector under all scenarios of performance is shown in Fig. 5.26. The power from the grid to the DES is shown by P_{12} and the power from CGS to DES is modeled by P_{45}. It is seen that all the flowed powers are associated with uncertainty except the power from the CGS to the DES (i.e., P_{45}). The CGS is equipped with the diesel generator and battery and both of them are designed to have a unique operating pattern under all scenarios of performance. In other words, they do not change their operating pattern together with changing uncertain parameters. As a result, the produced power by CGS (the diesel generator power and the discharged power from the battery) or the consumed power by CGS (charged power to the battery) are not affected by uncertainty and the CGS shows a unique pattern under all uncertainties. On the other hand, the uncertainties in the homes including load power uncertainty or renewable energy uncertainty are handled by grid power through P_{12}, P_{23}, and P_{34}.

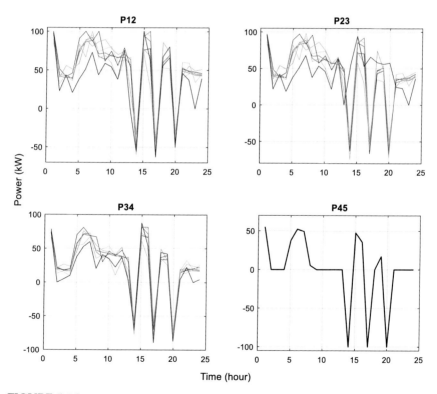

FIGURE 5.26   Flowed power in sectors of DES under all scenarios of performance.

In order to provide more details to the readers, the numerical values of powers in each sector are presented here. The power from the grid to the DES (P_{12}) under all scenarios of performance is presented in Table 5.32.

TABLE 5.32   Exchanged power per kW from node 1 to node 2 (P_{12}).

Hour	Scenario number						
	1	2	3	4	5	6	7
1	95.472	98.894	89.049	100	89.844	85.286	97.526
2	43.69	41.733	31.609	51.261	40.125	46.546	22.67
3	42.43	39.758	44.369	41.556	55.814	43.089	43.167
4	38.73	39.637	37.659	45.937	50.99	35.799	20.54
5	73.481	75.963	71.727	90.878	58.606	83.669	37.551
6	86.499	88.461	88.886	100	89.047	66.751	50.979
7	88.327	92.271	95.461	87.297	80.838	100	72.901
8	70.901	67.789	74.012	100	86.753	65.454	43.756
9	72.175	63.774	77.568	61.712	89.499	69.128	58.749
10	68.175	69.696	80.488	74.461	77.767	58.975	54.267
11	66.5	65.643	60.267	56.012	69.982	37.823	37.987
12	66.5	70.945	63.095	78.82	74.243	64.494	75.947
13	56.7	63.346	60.006	53.012	22.533	66.381	− 0.278
14	−51.15	−49.877	−33.3	−57.482	−33.994	−59.85	−55.572
15	94.701	93.493	100	76.108	76.967	71.743	99.937
16	87.176	100	94.087	77.913	67.986	29.251	62.312
17	−41.43	−43.736	−44.331	−57.777	−52.628	−42.183	−62.885
18	59.9	60.214	56.85	55.825	85.773	52.963	67.482
19	69.264	68.663	68.906	66.037	73.215	52.087	80.177
20	−44.14	−43.792	−34.347	−50.362	−31.616	−44.657	−44.727
21	53.21	49.13	39.594	51.466	43.342	60.402	45.929
22	48.17	46.506	45.24	48.937	55.351	59.361	37.433
23	45.15	43.559	46.937	47.676	49.19	34.268	0
24	43.75	42.776	44.19	45.84	51.236	45.156	39.237

The power from node 2 to node 3 (P_{23}) and power from node 3 to node 4 (P_{34}) under all scenarios of performance are listed in Tables 5.33 and 5.34. The power from DES to CGS (P_{45}) is summarized in Table 5.35.

TABLE 5.33 Exchanged power per kW from node 2 to node 3 (P_{23}).

Hour	Scenario number						
	1	2	3	4	5	6	7
1	92.972	96.402	86.533	97.138	86.786	83.415	95.116
2	41.19	39.293	28.996	48.669	35.734	43.841	21.808
3	39.93	37.34	42.486	38.792	54.119	39.838	42.137
4	37.23	38.163	36.497	44.063	49.16	34.489	18.554
5	71.981	74.302	70.743	88.814	57.185	81.744	37.551
6	83.099	84.772	85.954	97.165	85.971	61.738	50.979
7	84.827	88.629	91.726	83.347	75.227	98.023	67.773
8	66.901	63.997	70.285	95.679	80.237	59.488	37.882
9	67.55	58.315	72.429	57.451	86.192	63.811	53.431
10	61.8	63.448	73.63	67.443	74.812	50.618	46.008
11	57.5	56.322	52.728	48.116	56.666	28.297	21.754
12	56	59.714	51.782	64.428	65.879	56.278	63.359
13	44.7	49.577	47.192	45.57	16.767	36.388	0.06
14	−62.15	−62.138	−45.529	−63.095	−51.881	−74.819	55.677
15	85.201	84.46	90.192	71.112	72.483	58.196	94.104
16	80.176	92.041	85.617	70.363	58.001	30.813	53.15
17	−49.33	−51.082	−53.48	−66.48	−61.285	−52.411	65.842
18	46.4	47.657	41.127	43.082	74.371	34.984	58.36
19	51.164	50.596	54.467	47.241	48.292	32.168	55.973
20	−63.84	−63.46	−57.012	−69.652	−50.017	−72.064	57.594
21	36.21	31.057	27.148	34.161	25.45	42.022	24.788
22	35.17	33.774	30.999	36.788	36.853	43.874	22.988
23	36.15	33.893	36.63	37.521	40.34	26.273	0
24	39.25	38.658	38.788	42.569	45.216	42.235	36.435

The load curtailment is depicted in Fig. 5.27 and it is clear that the load is only curtailed under scenarios 6 and 7. In scenario 6, the loads of homes 2 to 6 are curtailed at hour 16. In scenario 7, the homes face load shedding only at hour 13. Homes 3 and 6 are exposed to the most load shedding because of heavy loads.

TABLE 5.34 Exchanged power per kW from node 3 to node 4 (P_{34}).

Hour	Scenario number						
	1	2	3	4	5	6	7
1	74.532	76.38	73.144	78.382	66.529	64.691	73.45
2	19.25	20.754	12.847	21.796	21.785	17.737	0
3	18.25	17.202	16.628	17.873	26.452	14.99	4.706
4	17.25	17.431	14.656	19.863	23.64	9.614	9.921
5	56.801	57.415	54.056	69.983	45.474	53.874	37.551
6	70.699	71.09	69.665	80.802	75.795	56.013	52.449
7	68.607	72.419	72.746	68.669	67.908	67.012	59.792
8	46.861	44.719	46.783	66.96	54.382	35.153	20.256
9	44.25	36.791	49.229	30.197	46.742	40.663	40.547
10	42.5	41.621	53.091	45.942	44.912	34.216	34.709
11	41	38.547	39.916	39.37	40.795	10.608	22.399
12	38.5	37.736	35.359	47.247	51.789	31.715	36.913
13	34	36.312	37.197	29.711	2.945	18.974	3.503
14	−70.25	−71.443	−57.469	−70.448	−58.079	−77.209	−67.085
15	81.701	81.499	82.994	69.148	71.997	61.36	87.319
16	71.376	81.069	77.609	66.063	66.741	35.454	53.799
17	−60.45	−63.05	−65.248	−69.962	−79.324	−57.464	−89.227
18	36.5	36.989	30.11	33.695	48.934	35.685	44.135
19	40.124	38.867	40.961	33.432	32.708	25.656	41.942
20	77.6	−79.107	−77.501	−82.185	−71.095	−81.139	−86.896
21	19.25	17.027	12.179	14.152	13.167	17.569	4.407
22	20.25	17.35	14.107	18.597	27.006	32.301	22.359
23	19.25	17.514	19.062	20.251	23.966	19.01	0
24	18.25	16.824	20.212	22.606	22.088	25.064	4.328

TABLE 5.35 Exchanged power per kW from node 4 to node 5 (P_{45}).

Hour	Power from DES to CGS (kW)
1	55.282
2	0
3	0
4	0
5	37.551
6	52.449
7	49.357
8	5.361
9	0
10	0
11	0
12	0
13	0
14	−100
15	47.701
16	35.376
17	−100
18	0
19	16.924
20	−100
21	0
22	0
23	0
24	0

In the CGS, the battery is operated as depicted in Fig. 5.28. The battery shows a unique operating regime under all uncertainties. It is charged in the initial hours of the day and the stored energy is restored and used at hours 14, 17, and 20.

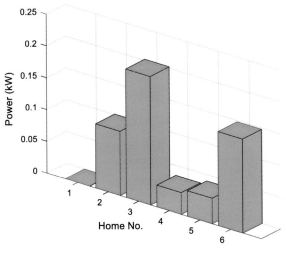

(A) hour 16-scenario 6

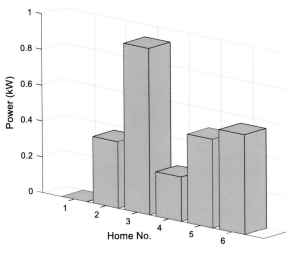

(B) hour 13-scenario 7

FIGURE 5.27   Load curtailment under scenarios 6 and 7.

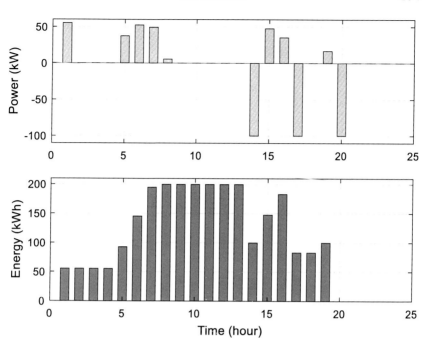

FIGURE 5.28   The operating pattern of battery under all scenarios of performance.

5.14 Conclusions

In this chapter, a typical DES connected to the external grid and CGS was modeled and simulated. The DES was supplied from two directions including grid and CGS. The DES was formed by 6 individual buildings with different loads and resources. Homes 1 and 3 acted as the prosumers and produced power by internal solar panels. Home 5 was another prosumer and produced energy via wind turbines. Each home supplied critical and noncritical loads. The DES was obligated to supply the critical loads under all events and uncertainties. The noncritical load could be curtailed or interrupted by DES when required. The CGS was integrated with the diesel generator and the battery energy storage system. The energy management system determined the optimal operating scheduling for the battery and the diesel generator, optimal load and energy management in each home, optimal energy exchange with the external grid and the CGS, and optimal operation for the p2p energy exchange.

In normal conditions when the DES has not faced an outage, the CGS and the grid supplied the homes. The diesel generator, p2p operation, and load curtailment were not used by the DES. At some hours like hour 1, all the energy of DES was supplied by the grid, and some energy was also injected to the CGS for charging the battery. At some other hours like hour 13, all the energy of homes in the DES was supplied by the CGS and the energy was also injected to the grid. Homes 1 and 3 produced power from their solar panels but their energy was not adequate to supply the loads and they needed to receive energy from the DES. However, home 5 was able to supply its loads as well as inject the surplus of wind energy to the DES at hours 14−19.

When line 23 was subjected to an outage in hours 11−15, the DES was supplied from both directions by the grid and the CGS. Homes 1 and 4 were supplied by the grid, and homes 2, 3, 5, and 6 were supplied by the CGS. The DES utilized the peer-to-peer operation and daily energy cost increased to 161 $/day. When line 34 faced an outage in hours 16−19, the DES was supplied from two directions by the grid and the CGS. The diesel generator and load curtailment were not utilized and the DES handled the outage by proper utilization of the battery. The daily energy cost increased to 156.68 $/day. During the outage of line 45 in hours 6−17, the DES was only supplied by the grid. The CGS was disconnected from DES and did not participate in the energy management system. The grid supplied all homes. The diesel generator, p2p operation, and load curtailment were not utilized. The daily energy cost increased to 157.53 $/day.

The outage of line 12 in hours 14 to 24 formed off-grid DES and all homes were supplied by CGS. The diesel generator and load curtailment were utilized and such operating increased the daily energy cost to 273.367 $/day. The load curtailment was applied on the noncritical loads in all homes and the critical loads were supplied at all hours. When line 12 faced an outage for the whole day, the DES needed to be supplied by the CGS for 24-hour resulting in more operation of the diesel generator. The load curtailment was applied to switch off the noncritical loads. Such points increased the daily energy cost to 820.800 $/day. When home 5 was disconnected from the DES, the peer-to-peer operation was used by the DES to supply loads in home 5. In the stochastic model, the load power, wind energy, and solar energy were modeled by a set of scenarios. The uncertainty increased the daily energy cost by 6%. The diesel generator and battery in CGS were designed for having a unique operating pattern under all scenarios of performance. The uncertainties in loads and renewables were handled by grid power.

References

[1] http://www.energy.gov. TUSDoE.
[2] Huang P, Copertaro B, Zhang X, Shen J, Löfgren I, Rönnelid M, et al. A review of data centers as prosumers in district energy systems: renewable energy integration and waste heat reuse for district heating. Applied Energy 2020;258:114109.

[3] Lund H, Østergaard PA, Chang M, Werner S, Svendsen S, Sorknæs P, et al. The status of 4th generation district heating: research and results. Energy 2018;164:147−59.
[4] Faraji H, Hemmati R. A novel resilient concept for district energy system based on central battery and decentral hybrid generating resources. International Journal of Energy Research 2022;46(9):11925−42.
[5] Mehrjerdi H, Mahdavi S, Hemmati R. Resilience maximization through mobile battery storage and diesel DG in integrated electrical and heating networks. Energy 2021;237:121195.
[6] Sabouhi H, Doroudi A, Fotuhi-Firuzabad M, Bashiri M. Electrical power system resilience assessment: a comprehensive approach. IEEE Systems Journal 2020;14(2):2643−52.
[7] Hemmati R, Mehrjerdi H, Nosratabadi SM. Resilience-oriented adaptable microgrid formation in integrated electricity-gas system with deployment of multiple energy hubs. Sustainable Cities and Society 2021;71:102946.
[8] Mehrjerdi H, Hemmati R. Coordination of vehicle-to-home and renewable capacity resources for energy management in resilience and self-healing building. Renewable Energy 2020;146:568−79.
[9] Wang Y, Rousis AO, Strbac G. Resilience-driven optimal sizing and pre-positioning of mobile energy storage systems in decentralized networked microgrids. Applied Energy 2022;305:117921.
[10] Hemmati R. Mobile model for distributed generations and battery energy storage systems in radial grids. Journal of Renewable and Sustainable Energy 2019;11(2):025301.
[11] Rahimi K, Davoudi M. Electric vehicles for improving resilience of distribution systems. Sustainable Cities and Society 2018;36:246−56.
[12] Aljabery AAM, Mehrjerdi H, Mahdavi S, Hemmati R. Multi carrier energy systems and energy hubs: comprehensive review, survey and recommendations. International Journal of Hydrogen Energy 2021;46(46):23795−814.
[13] Mehrjerdi H. Peer-to-peer home energy management incorporating hydrogen storage system and solar generating units. Renewable Energy 2020;.
[14] Nosratabadi SM, Hemmati R, Khajouei Gharaei P. Optimal planning of multi-energy microgrid with different energy storages and demand responsive loads utilizing a technical-economic-environmental programming. International Journal of Energy Research 2021;45(5):6985−7017.
[15] Mehrjerdi H, Hemmati R, Mahdavi S, Shafie-khah M, Catalao JP. Multi-carrier microgrid operation model using stochastic mixed integer linear programming. IEEE Transactions on Industrial Informatics 2022;18(7):4674−87.
[16] Hemmati R. Stochastic energy investment in off-grid renewable energy hub for autonomous building. IET Renewable Power Generation 2019;13(12):2232−9.

Smart homes and microgrids on the electric distribution grids

Energy Management in Homes and Residential Microgrids
DOI: https://doi.org/10.1016/B978-0-443-23728-7.00006-0

261

© 2024 Elsevier Inc. All rights reserved.

Nomenclature

Indexes and sets

i, j	Index of buses in the distribution grid
s	Index of scenarios of performance
t	Index of time periods
N	Set of buses in the distribution grid
Nn	Set of buses without smart homes and microgrids
Nh	Set of buses with smart homes
Nm	Set of buses with smart microgrids
S	Set of scenarios of performance
T	Set of time periods

Parameters and variables

C_{ele}^t	Electricity price (\$/kWh)
C_f	Fuel price (\$/kWh)
d_{ti}^t	Duration of time interval (hour)
$E_b^{i,t}$	Energy stored by battery (kWh)
$E_b^{i,0}$	Initial energy inside battery (kWh)
$E_b^{i,r}$	Rated capacity of battery (kWh)
k_{pr}	Factor for converting per-unit value to real value
k_{rp}	Factor for converting real value to per-unit value
O_{pr}^s	Probability of occurrence for each scenario of performance
$P_b^{i,r}$	Rated power of battery (kW)
$P_{cb}^{i,t}$	Charged power to battery (kW)
$P_{db}^{i,t}$	Discharged power from battery (kW)
$P_{dg}^{i,t}$	Power of diesel generator (kW)
$P_{dg}^{i,r}$	Rated power of diesel generator (kW)
$P_f^{1,2,t}$	Flowed power in line between bus 1 and bus 2 (p.u.)
$P_f^{1,2,t,s}$	Flowed power in line between bus 1 and bus 2 under scenarios of performance (p.u.)
$P_f^{i,j,t}$	Flowed power in line between bus i and bus j (p.u.)
$P_f^{i,j,r}$	Rated power of line between bus i and bus j (p.u.)
$P_h^{i,t}$	Consumed power by smart home (kW)
$P_l^{i,t}$	Consumed power by load (kW)
$P_m^{i,t}$	Consumed power by smart microgrid (kW)
$P_m^{i,r}$	Rated power of line between microgrid and distribution grid (kW)
$P_s^{i,t}$	Produced power by solar generating system (kW)
$P_w^{i,t}$	Produced power by wind generating system (kW)
u_{cb}^i	Binary variable showing charging state of battery
u_{db}^i	Binary variable showing discharging state of battery
$y^{i,j}$	Susceptance of line between bus i and bus j (p.u.)
z_{dg}	Daily operating cost of diesel generators (\$/day)
z_{ele}	Daily cost of purchasing electricity by distribution grid (\$/day)
z_{daily}	Daily energy cost of distribution grid (\$/day)
$\theta^{i,t}$	Voltage angle (Radian)
η_b^i	Efficiency of battery (%)

6.1 Introduction

In the regular distribution grids, all the loads are supplied by the grid and the energy management system can only be carried out on the loads for load regulation and management. In such systems, the loads may be curtailed, interrupted, or shifted to deal with the matters of the distribution grid such as voltage issues, energy losses, congestion of lines, and on-peak loading. On the other hand in the active distribution grids, the distributed energy resources (DERs) or distributed generations (DGs) are integrated into the distribution grid and these local resources can participate in the energy management system [1]. The DGs may be nonrenewable resources such as diesel generators and microgas turbines [2], or they could be renewable resources such as wind and solar generating systems [3]. The energy storage systems like battery and hydrogen storage are as well integrated into the distribution networks for forming active distribution grids [4,5]. Emerging technologies such as electric vehicles and their charging stations can also be integrated into the distribution networks for creating active distribution grids [6]. All of those local resources make significant impacts on both the technical and economic characteristics of the distribution grid. In such grids, the energy management system should optimize the operating patterns of resources in order to achieve specific purposes such as minimum energy cost, voltage regulation, and energy loss reduction. The local resources may be owned by private companies and managed for maximizing profit [5,7].

One of the significant advantages of local resources on the distribution grid is dealing with power outages and increasing the resilience of the grid [8]. When the distribution grid faces a blackout, the local resources are able to supply some parts of loads or even all of them on the distribution grid. The energy storage systems and DGs can be efficiently managed for improving the resilience of the grid [9]. Those local resources enable the distribution grid to continue operation off-grid when necessary [10]. Grid formation is one of the achievements of installing local resources on the distribution grid. In the grid formation, the distribution grid is divided into several subsections and each subsection can continue operating separately. Such operation is very useful when some subsections are faced with issues because the healthy subsections are not affected by the issues of faulty subsections [11].

Apart from the DGs, the distribution grids may be integrated with smart homes [12] and microgrids [13]. In such a paradigm, each home or microgrid can comprise several DGs at the same time. All the resources in the smart homes and microgrids need to be optimally operated by the energy management system. In such systems, the local homes and microgrids enable the distribution grid for off-grid operation when the upstream network is faced with a blackout [14].

This chapter models, simulates and discusses the integration of microgrids and smart homes into the active distribution grid. In normal operating conditions, the smart homes and microgrids are managed and their operations are optimized by the energy management system in order to minimize the energy cost. In the off-grid operation of feeders, the smart homes and microgrids are managed to supply the loads on the off-grid feeders. The main highlights of this chapter are summarized as follows:

- The active distribution grid is formed with two feeders. The first feeder is integrated with three smart homes on buses 8, 10, and 18, and one microgrid on bus 13. The second feeder is supplied by one microgrid on bus 30, and one smart home on bus 32.
- Each microgrid is equipped with a diesel generator, wind generating system, solar generating system, and battery energy storage system. The energy management system optimizes the operating pattern of the diesel generator and the charging scheduling of the battery. The maximum power is extracted from the wind and solar systems as well.
- Smart homes are supplied by wind and solar generating systems. Since these homes are not equipped with an energy storage system, the energy management system is not able to regulate the power energy in the smart homes and only extracts the maximum power from wind and solar systems.
- In the normal operating condition, the energy management system minimizes the energy cost of the active distribution grid by optimal operation of microgrids and extracting maximum power from renewable resources in smart homes.
- In the off-grid operation of feeders, the energy management system utilizes local resources on each feeder for supplying the loads. As well, the operating cost of diesel generators is minimized.
- The uncertainty of load, wind, and solar powers is modeled by means of stochastic programming.

6.1.1 Some research gaps in the integration of homes and microgrids into the grid

There are some items that have not been sufficiently addressed by researchers or have been rarely and marginally investigated in the literature but those items are important and they need to be extended in future studies. The outlines of some major points are summarized here.

1. **Multimicrogrids and multihomes on the distribution grid:** The multimicrogrids and multihomes may be integrated into the

distribution grid at the same time. Those systems are owned by private participants and their integration into the grid should be managed subject to the satisfaction and profit of users. The energy exchange between microgrids and homes may be done based on the energy transaction market.

2. **Mobile resources:** Mobile energy storage systems and DGs such as truck-mounted battery energy storage systems and truck-mounted diesel generators may be regarded in the proposed paradigm. The mobile resources may be transferred daily, weekly, or even seasonally.

3. **Energy management considers a three-phase unbalanced power flow:** Some resources and loads on the distribution grid may operate as single-phase. They need to be modeled by a 3-phase 4-wire unbalanced power flow.

4. **Cloud energy storage systems and resources:** The cloud energy systems such as cloud energy storage and cloud energy resources present a concept opposed to conventional DERs. In the conventional DERs, each home is integrated with energy storage systems or renewable resources. In cloud energy systems, a large energy storage system or a big energy resource is equipped on the grid and it devotes adequate capacity to each home. The homes are not physically equipped with energy resources and they use the capacity of the cloud energy system. The cloud energy resource is a shared DER that is capable of providing energy storage or energy-generating services at a substantially lower cost. The control system, communication model, and business model need to be developed for cloud energy systems.

6.2 Distribution grid with smart homes and microgrids

In the regular electric distribution grids, as shown in Fig. 6.1, the loads on the buses are supplied by the upstream transmission network [5]. The transmission network is connected to several distribution grids and supplies them. The loads on the buses of the distribution grid may be small loads such as homes and buildings or large loads like the university campus and the industrial districts. The distribution grids are formed by several feeders and each feeder usually supplies a different area. The distribution grid would be integrated with DG systems, microgrids, or smart homes. In such cases, the distribution grid is able to supply some parts of its loads or even can send power to the upstream transmission network by means of local resources. This type of system is called an active distribution grid. Fig. 6.2 shows the active distribution grid integrated with microgrids and smart homes on the feeders.

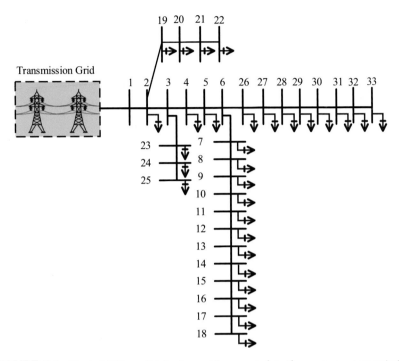

FIGURE 6.1 Typical 33-bus distribution grid connected to the upstream transmission network.

Feeder 1 is integrated with three smart homes on buses 8, 10, and 18 and one microgrid on bus 13. Feeder 2 is integrated with one microgrid on bus 30 and one smart home on bus 32. Microgrids and smart homes are able to exchange energy with the active distribution grid.

The active distribution grid with smart homes and microgrids can be designed properly to supply some loads when the upstream transmission network is not connected or is not available because of a power outage or a blackout. Fig. 6.3 demonstrates a typical formation of a grid, where the local resources including smart homes and microgrids supply feeders 1 and 2 and those two feeders are not connected to the active distribution grid. The off-grid operation of feeders 1 and 2 is feasible through proper control and management of energy resources in smart homes and microgrids. In this situation, the renewable resources, energy storage systems, and nonrenewable resources in the smart homes and microgrids are managed by the energy management system to deal with loading on all buses of the off-grid district.

In the proposed test system, each microgrid is equipped with a diesel generator, wind generating system, solar generating system, and battery energy storage. The smart home on bus 8 is supplied by both wind and

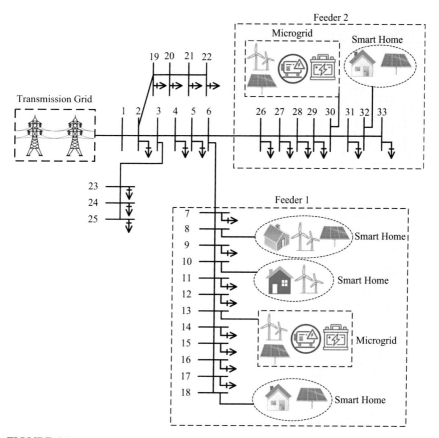

FIGURE 6.2 Distribution grid integrated with smart homes and microgrids on the feeders.

solar generating systems. The smart home on bus 10 is integrated with the wind-generating system. The smart homes on buses 18 and 32 are equipped with solar-generating systems.

6.3 Formulation and modeling

The active distribution grid with smart homes and microgrids is modeled to minimize energy costs when feeders 1 and 2 are connected to the grid. When the feeders are disconnected and operate as off-grid or islanded, the energy management system utilizes the resources on each feeder to supply the loads. Additionally, the operating cost of resources such as fuel cost for nonrenewable resources is minimized. The cost of purchased electricity from the upstream transmission network is calculated by Eq. (6.1). This cost is related to the loads on all

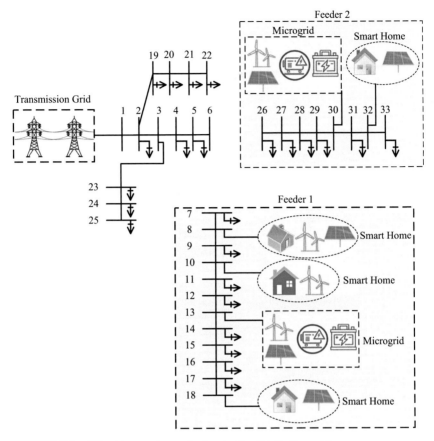

FIGURE 6.3   Off-grid operation on feeders in the distribution grid.

buses of the active distribution grid. While the upstream transmission network is available, the distribution grid prefers to receive the required energy of loads from the transmission network rather than utilizing nonrenewable resources like diesel generators [15]. However, at some hours it would be required to utilize the diesel generators and the related operating cost is calculated by Eq. (6.2). The total energy cost of the distribution grid is achieved as Eq. (6.3) which is the sum of the above-mentioned costs [5].

$$z_{ele} = \sum_{t=1}^{T} \left(P_f^{1,2,t} \times C_{ele}^t \times k_{pr} \right) \tag{6.1}$$

$$z_{dg} = \left\{ \sum_{i \in Nm}^{N} \left(\sum_{t=1}^{T} \left(P_{dg}^{i,t} \right) \right) \right\} \times C_f \tag{6.2}$$

$$z_{daily} = z_{ele} + z_{dg} \tag{6.3}$$

The power flow in the distribution grid needs to be verified under any operating conditions of smart homes and microgrids. The flowed power through each line of the distribution grid is formulated as Eq. (6.4). The flowed power through each line is restricted by the rated capacity of the line as expressed by Eq. (6.5) [15].

$$
\begin{aligned}
P_f^{i,j,t} &= y^{i,j} \times \left(\theta^{i,t} - \theta^{j,t} \right) \\
&\forall i \in N \\
&\forall j \in N \\
&\forall t \in T
\end{aligned}
\tag{6.4}
$$

$$
\begin{aligned}
\left| P_f^{i,j,t} \right| &\leq P_f^{i,j,r} \\
&\forall i \in N \\
&\forall j \in N \\
&\forall t \in T
\end{aligned}
\tag{6.5}
$$

The power balance on all buses of the distribution grid needs to be verified. There are three sets of buses including the buses without smart homes and microgrids, the buses with smart homes, and the buses with microgrids. The power balance on the buses without smart homes and microgrids is specified by Eq. (6.6). The power equilibrium on the buses with smart homes is formulated as Eq. (6.7) and the power balance on the buses integrated with microgrids is expressed by Eq. (6.8) [15].

$$
\begin{aligned}
P_l^{i,t} \times k_{rp} + \sum_{j=1}^{N} P_f^{i,j,t} &= 0 \\
&\forall i \in Nn \\
&\forall t \in T
\end{aligned}
\tag{6.6}
$$

$$
\begin{aligned}
P_h^{i,t} \times k_{rp} + \sum_{j=1}^{N} P_f^{i,j,t} &= 0 \\
&\forall i \in Nh \\
&\forall t \in T
\end{aligned}
\tag{6.7}
$$

$$
\begin{aligned}
P_m^{i,t} \times k_{rp} + \sum_{j=1}^{N} P_f^{i,j,t} &= 0 \\
&\forall i \in Nm \\
&\forall t \in T
\end{aligned}
\tag{6.8}
$$

Every smart home exchanges power with the distribution grid as indicated by Eq. (6.9). The produced power by local resources is used to supply the local loads in the homes and the surplus of energy can be injected into the distribution grid. If the local resources are not adequate to supply the local loads, the home may receive power from the distribution grid. At some hours, the produced energy by local resources is exactly equal to the consumed energy by local loads and there is not any shortage or surplus of energy in the home. In such time periods, the smart home neither injects power into the grid nor takes power from it. A similar operation is defined for microgrids as expressed by Eq. (6.10). In the microgrids, the larger resources are equipped. As well, some extra resources like bulk energy storage systems are available. The microgrids exchange power with the distribution grid and are able to send power to the grid or take power from it. The capacity of the interconnecting line between the microgrid and the distribution grid is modeled by Eq. (6.11) [15].

$$P_h^{i,t} = \left\{ P_l^{i,t} \right\} - \left\{ P_s^{i,t} + P_w^{i,t} \right\}$$
$$\forall i \in Nh$$
$$\forall t \in T$$
(6.9)

$$P_m^{i,t} = \left\{ P_l^{i,t} + P_{cb}^{i,t} \right\} - \left\{ P_{db}^{i,t} + P_s^{i,t} + P_w^{i,t} + P_{dg}^{i,t} \right\}$$
$$\forall i \in Nm$$
$$\forall t \in T$$
(6.10)

$$\left| P_m^{i,t} \right| \leq P_m^{i,r}$$
$$\forall i \in Nm$$
$$\forall t \in T$$
(6.11)

The microgrids are integrated with wind turbines, solar panels, diesel generators, and battery energy storage systems. The operating pattern of the diesel generator is assumed as a design variable and optimized by the energy management system. The produced power by the diesel generator at each hour must be less than the rated capacity of the diesel generator as modeled by Eq. (6.12). The operating pattern (charging scheduling) of the battery energy storage system is regarded as a design variable and optimized by the energy management system. The charged power to the battery and the discharged power from the battery at each hour must be less than the rated power of the energy storage system as developed through Eqs. (6.13) to (6.15) [15].

$$\left| P_{dg}^{i,t} \right| \leq P_{dg}^{i,r}$$
$$\forall i \in Nm$$
$$\forall t \in T$$
(6.12)

$$u_{cb}^i + u_{db}^i \leq 1$$
$$\forall i \in Nm$$
$$\forall t \in T$$
(6.13)

$$P_{cb}^{i,t} \leq P_b^{i,r} \times u_{cb}^i$$
$$\forall i \in Nm$$
$$\forall t \in T$$
(6.14)

$$P_{db}^{i,t} \leq P_b^{i,r} \times u_{db}^i$$
$$\forall i \in Nm$$
$$\forall t \in T$$
(6.15)

The stored energy by battery energy storage system at each hour is formulated by Eq. (6.16). If the battery energy storage system contains initial energy at beginning of programming, it can be modeled by Eq. (6.17). The stored energy inside the battery at each hour must be less than the rated capacity of the battery as shown by Eq. (6.18). The battery efficiency is defined by Eq. (6.19).

$$E_b^{i,t} = E_b^{i,t-1} + \left(P_{cb}^{i,t} - \frac{P_{db}^{i,t}}{\eta_b^i} \right) \times d_{ti}^t$$
$$\forall i \in Nm$$
$$\forall t \in T, t \notin [1]$$
(6.16)

$$E_b^{i,t} = E_b^{i,0} + \left(P_{cb}^{i,t} - \frac{P_{db}^{i,t}}{\eta_b^i} \right) \times d_{ti}^t$$
$$\forall i \in Nm$$
$$\forall t \in [1]$$
(6.17)

$$E_b^{i,t} \leq E_b^{i,r}$$
$$\forall i \in Nm$$
$$\forall t \in T$$
(6.18)

$$\eta_b^i = \frac{\sum_{t=1}^{T} \left[P_{db}^{i,t} \times d_{ti}^t \right]}{\sum_{t=1}^{T} \left[P_{db}^{i,t} \times d_{ti}^t \right]}$$
$$\forall i \in Nm$$
(6.19)

6.4 Data of illustrative test case

The illustrative distribution grid is integrated with two microgrids on buses 13 and 30. Each microgrid comprises a diesel generator, wind-generating system, solar-generating system, and battery energy storage. The data of these resources of microgrids are listed in Table 6.1 [5].

TABLE 6.1 Data of microgrids on buses 13 and 30.

Parameter	Level
Rated power of microgrid line (MW)	1.5
Rated power of battery (MW)	0.4
Rated capacity of battery (MWh)	1.2
Initial energy of battery (MWh)	0
Rated power of diesel generator (MW)	0.5
Rated power of solar generating system (MW)	0.4
Rated power of wind generating system (MW)	0.4

TABLE 6.2 Data of smart homes on buses 8, 10, 18, and 32.

Parameter	Location	Level
Rated power of wind system (kW)	Home on bus 8	100
Rated power of solar system (kW)	Home on bus 8	100
Rated power of wind system (kW)	Home on bus 10	60
Rated power of solar system (kW)	Home on bus 18	90
Rated power of solar system (kW)	Home on bus 32	210

The fuel price for the operation of the diesel generator is equal to 0.15 $/kWh. The distribution grid is integrated with four smart homes on buses 8, 10, 18, and 32. The smart home on bus 8 is equipped with both wind and solar generating systems. The smart home on bus 10 is supplied by the wind generating system. The smart homes on buses 18 and 32 are equipped with solar generating systems. The rated powers of wind and solar generating systems in these smart homes are presented in Table 6.2.

The distribution grid comprises 33 buses and 32 lines. The capacity and susceptance of each line are presented in Table 6.3. The data are presented as per-unit. In the per-unit system, the base apparent power is 10 MVA and the base voltage is 12.66 kV. The line numbers are defined according to Fig. 6.2, where, feeder 1 includes lines 7–17 and feeder 2 comprises lines 26–32. The loads on the buses of the distribution grid are depicted in Fig. 6.4. Bus 1 is not integrated with load and the largest loads are on buses 23–25. Table 6.4 presents the daily power profiles of loads and renewable resources. The electricity price over 24-hour is depicted in Fig. 6.5.

TABLE 6.3   The 33-bus distribution grid parameters.

Line Number	From bus	To bus	Susceptance (p.u.)	Line capacity (p.u.)
1	1	2	−341.011	0.8532
2	2	3	−63.8294	0.7498
3	3	4	−85.9848	0.5152
4	4	5	−82.5737	0.4876
5	5	6	−22.6698	0.4738
6	6	7	−25.9010	0.2484
7	7	8	−68.1734	0.2000
8	8	9	−21.6589	0.1540
9	9	10	−21.6589	0.1402
10	10	11	−246.577	0.1266
11	11	12	−129.463	0.1172
12	12	13	−13.8767	0.1034
13	13	14	−22.4822	0.0828
14	14	15	−30.4706	0.0896
15	15	16	−29.4084	0.0620
16	16	17	−9.3129	0.0484
17	17	18	−27.9226	0.0206
18	2	19	−102.412	0.0828
19	19	20	−11.8250	0.0620
20	20	21	−33.5024	0.0414
21	21	22	−17.0997	0.0206
22	3	23	−51.9869	0.2138
23	23	24	−22.6027	0.1932
24	24	25	−22.8606	0.0966
25	6	26	−155.005	0.2116
26	26	27	−110.764	0.1978
27	27	28	−17.1656	0.184
28	28	29	−22.8769	0.1702
29	29	30	−62.0022	0.1426
30	30	31	−16.6434	0.0966
31	31	32	−44.2873	0.0620
32	32	33	−30.2293	0.0138

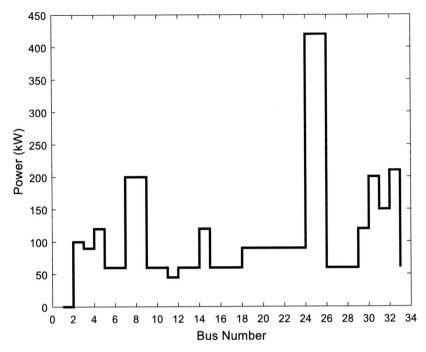

FIGURE 6.4 Loading on buses of distribution grid.

TABLE 6.4 Daily power profiles of loads and renewable.

Hour	Load (%)	Solar (%)	Wind (%)
1	25	0	10
2	20	0	56
3	20	0	82
4	15	0	52
5	20	0	44
6	25	0	70
7	40	10	76
8	60	15	82
9	75	30	84
10	75	60	84
11	80	80	100
12	80	100	100

(Continued)

Energy Management in Homes and Residential Microgrids

TABLE 6.4 (Continued)

Hour	Load (%)	Solar (%)	Wind (%)
13	85	90	78
14	80	85	64
15	70	70	100
16	60	50	92
17	70	30	84
18	80	15	80
19	90	10	78
20	100	5	50
21	90	0	40
22	75	0	10
23	55	0	10
24	35	0	10

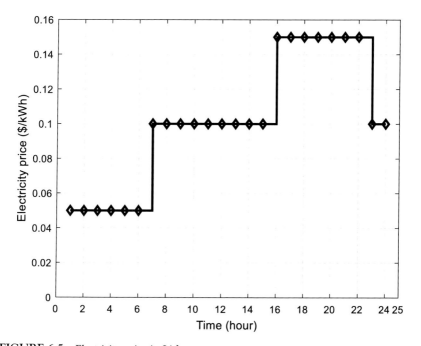

FIGURE 6.5 Electricity price in 24-hour.

6.5 Normal operation of feeder 1

In normal operating conditions, the smart homes and microgrids on feeder 1 participate in the energy management system as presented in Table 6.5. The smart homes are only equipped with wind and solar generating systems. Those homes use their renewable energy to supply their loads and the surplus of generated energy is sent to the distribution grid. The smart home on bus 8 injects power to the distribution grid at many hours like 2–7, 11–12, and 15–16. In the rest of the hours,

TABLE 6.5 Produced-consumed power by smart homes and microgrid on feeder 1.

Hour	Smart home on bus 8 (kW)	Smart home on bus 10 (kW)	Microgrid on bus 13 (kW)	Smart home on bus 18 (kW)
1	40	9	−25	22.5
2	−16	−21.6	−212	18
3	−42	−37.2	84	18
4	−22	−22.2	201	13.5
5	−4	−14.4	236	18
6	−20	−27	−265	22.5
7	−6	−21.6	−320	27
8	23	−13.2	−352	40.5
9	36	−5.4	−811	40.5
10	6	−5.4	−531	13.5
11	−20	−12	−672	0
12	−40	−12	−1152	−18
13	2	4.2	−221	−4.5
14	11	9.6	−148	−4.5
15	−30	−18	−638	0
16	−22	−19.2	−532	9
17	26	−8.4	−414	36
18	65	0	−732	58.5
19	92	7.2	−698	72
20	145	30	−160	85.5
21	140	30	−506	81
22	140	39	5	67.5
23	100	27	−7	49.5
24	60	15	−19	31.5

the power is taken from the grid. The smart home on bus 10 sends power to the grid for many hours of the day. However, the exchanged power between the smart home on bus 18 and the distribution grid is mostly positive which means flowing power from the grid toward the home. It confirms that this smart home is not able to send large amounts of power to the grid and mostly receives power from the distribution network. In this smart home, the produced energy by local renewable resources is used to supply the loads and there is not too much excess energy that could be exchanged with the distribution grid. The microgrid on bus 13 efficiently participates in the energy management system. There is a difference between the operations of microgrids and smart homes. The smart homes are only equipped with renewable resources while the microgrid is equipped with renewable resources, nonrenewable resources, and energy storage systems. In renewable resources, the produced power at each hour cannot be regulated or managed and all of the produced power must be either used by the home or injected into the grid. As a result, the energy management system cannot regulate or manage the exchanged energy between the smart homes and the distribution grid. In such homes, the energy management system attempts to harvest maximum power from renewable

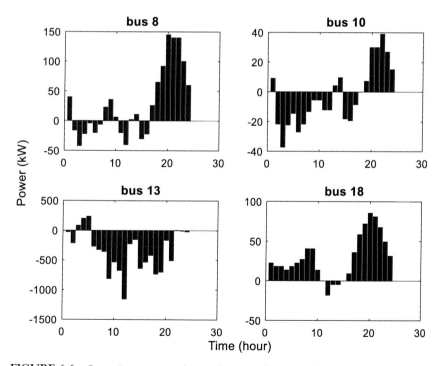

FIGURE 6.6   Operating patterns of smart homes and microgrid on feeder 1.

resources to prevent renewable energy waste. On the other hand, in the microgrids, the maximum power is extracted from renewable resources but the produced power by nonrenewable resources can be managed or regulated. As well, the charging scheduling of the energy storage system can be managed. Therefore, the energy management system can efficiently regulate or manage the exchanged energy between the microgrid

TABLE 6.6 Flowed power per kW in lines of feeder 1.

Hour	Line number										
	7	8	9	10	11	12	13	14	15	16	17
1	162.75	122.75	107.75	98.75	87.5	72.5	97.5	67.5	52.5	37.5	22.5
2	−138.6	−122.6	−134.6	−113	−122	−134	78	54	42	30	18
3	115.8	157.8	145.8	183	174	162	78	54	42	30	18
4	240.05	262.05	253.05	275.25	268.5	259.5	58.5	40.5	31.5	22.5	13.5
5	328.6	332.6	320.6	335	326	314	78	54	42	30	18
6	−173.25	−153.25	−168.25	−141.25	−152.5	−167.5	97.5	67.5	52.5	37.5	22.5
7	−134.6	−128.6	−152.6	−131	−149	−173	147	99	75	51	27
8	−22.7	−45.7	−81.7	−68.5	−95.5	−131.5	220.5	148.5	112.5	76.5	40.5
9	−391.15	−427.15	−472.15	−466.75	−500.5	−545.5	265.5	175.5	130.5	85.5	40.5
10	−168.15	−174.15	−219.15	−213.75	−247.5	−292.5	238.5	148.5	103.5	58.5	13.5
11	−332	−312	−360	−348	−384	−432	240	144	96	48	0
12	−850	−810	−858	−846	−882	−930	222	126	78	30	−18
13	175.95	173.95	122.95	118.75	80.5	29.5	250.5	148.5	97.5	46.5	−4.5
14	240.1	229.1	181.1	171.5	135.5	87.5	235.5	139.5	91.5	43.5	−4.5
15	−360.5	−330.5	−372.5	−354.5	−386	−428	210	126	84	42	0
16	−285.2	−263.2	−299.2	−280	−307	−343	189	117	81	45	9
17	−34.9	−60.9	−102.9	−94.5	−126	−168	246	162	120	78	36
18	−236.5	−301.5	−349.5	−349.5	−385.5	−433.5	298.5	202.5	154.5	106.5	58.5
19	−108.3	−200.3	−254.3	−261.5	−302	−356	342	234	180	126	72
20	565.5	420.5	360.5	330.5	285.5	225.5	385.5	265.5	205.5	145.5	85.5
21	163.5	23.5	−30.5	−60.5	−101	−155	351	243	189	135	81
22	600.25	460.25	415.25	376.25	342.5	297.5	292.5	202.5	157.5	112.5	67.5
23	425.25	325.25	292.25	265.25	240.5	207.5	214.5	148.5	115.5	82.5	49.5
24	250.25	190.25	169.25	154.25	138.5	117.5	136.5	94.5	73.5	52.5	31.5

and the distribution grid. The produced-generated power by microgrid confirms such matter. The microgrid sends power to the distribution grid during most of the hours and receives power from the grid during low-priced energy hours.

Fig. 6.6 shows the operating patterns of the smart homes and the microgrid on Feeder 1. The positive values show the power from the grid to the home-microgrid and the negative values show power from the home-microgrid toward the grid. It is clear that the home on bus 18 sends minimum power to the grid and the microgrid on bus 13 injects maximum power to the grid.

Table 6.6 presents the flowed power through lines of feeder 1. The line numbers are already defined in Table 6.3. This feeder takes power from the distribution grid at many hours but it is able to send power back to the distribution grid at hours like 6−12 and 15−19 especially when the energy is expensive. The microgrid is installed on bus 13 which is located between lines 12 and 13. It is seen that the microgrid produces power for many hours and injects it into lines 12 and 13. The injected power to line 13 is used to supply the loads on buses 14−18 (the buses that are located after bus 13). The injected power to line 12 is used to supply the loads on the buses that are located before bus 13 as well as for exchanging power with the transmission network.

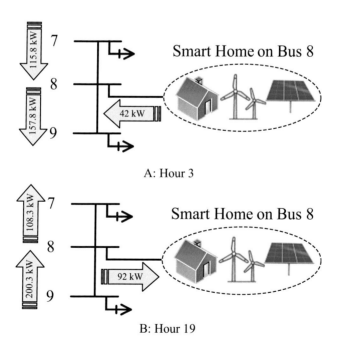

FIGURE 6.7   Power balance in the location of smart home on bus 8.

A significant part of the exchange power with the transmission network comes not from the smart homes but from the microgrids.

The power balance in the location of smart homes and microgrids is evaluated here. The power balance in the location of the smart home on bus 8 is shown in Fig. 6.7. At hour 3, the smart home produces power and its power is added to the power from the grid for supplying the loads on the buses located after bus 8 onward. At hour 19, the power direction is from the feeder toward the distribution grid. The smart home consumes some part of the energy and the rest of the energy is forwarded to the buses located before bus 8.

The power balance in the location of the smart home on bus 10 is depicted in Fig. 6.8. The smart home produces power at hour 2. At hour 18, the exchanged power between the smart home and the grid is zero. In other words, the produced power by renewable resources in the

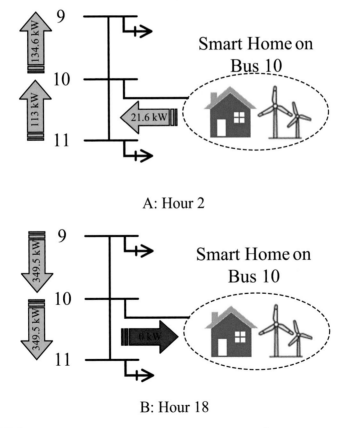

FIGURE 6.8   Power balance in the location of smart home on bus 10.

home is adequate to supply its loads. As a result, the home neither takes energy from the grid nor sends energy to the grid.

The microgrid operation at hours 1, 3, and 19 is depicted in Fig. 6.9. At hour 1, the microgrid produces 25 kW power and its power is added

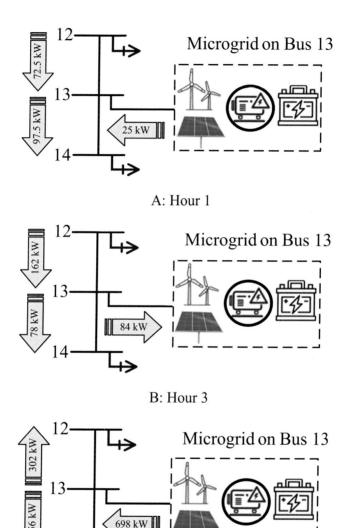

A: Hour 1

B: Hour 3

C: Hour 19

FIGURE 6.9   Power balance in the location of microgrid on bus 13.

to the power from the grid for supplying loads of feeder 1. At hour 2, the distribution grid is still supplying loads of feeder 1 as well as loads of the microgrid. At hour 19, the microgrid is supplying both the distribution grid and loads of feeder 1. The microgrid produces 356 kW to supply the loads on the buses after bus 13 as well as it produces 302 kW to supply the loads on the buses before bus 13.

The power balance in the location of the smart home on bus 18 is shown in Fig. 6.10. The smart home consumes energy at hour 1 and

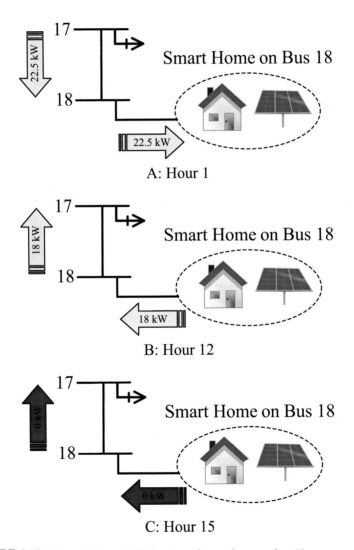

FIGURE 6.10   Power balance in the location of smart home on bus 18.

produces power at hour 12. The exchanged power between the smart home and the grid is zero at hour 15 and the smart home operates off-grid at this hour.

As already pointed out, the microgrid is installed on bus 13 which is located between lines 12 and 13. The flowed power through these two lines (i.e., the line from bus 12 to 13 and the line from bus 13 to 14) is depicted in Fig. 6.11. The negative powers in the line from bus 12 to 13 mean that the direction of power is from bus 13 to 12. It is clear that the microgrid sends power back to the grid through the line from bus 12 to 13. This power may be consumed by the loads that are located before bus 13 or may be traded with the upstream transmission network. The energy management system optimizes the operation by deciding on this subject. The power from bus 13 to 14 is always positive and it confirms that the smart homes in the locations after bus 13 cannot send power back to the distribution grid.

Fig. 6.12 shows the operating pattern of the battery in microgrid 1. The battery charges energy at hours 3–5 and discharges it at hours 9 and 12. As well, it is charged again at hours 13–14 and discharged at hours 18–19 and 21. It is clear that the energy is stored when it is low-priced and the stored energy is used when the cost of energy is high

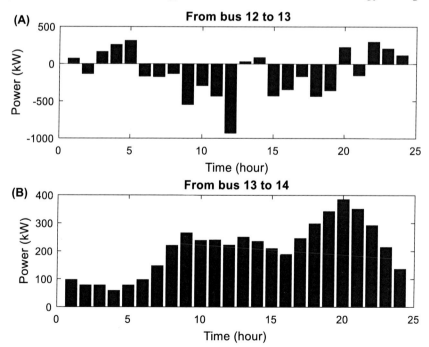

FIGURE 6.11 Flowed power in lines of distribution grid connected to bus 13. (A) Line from bus 12 to 13; (B) Line from bus 13 to 14.

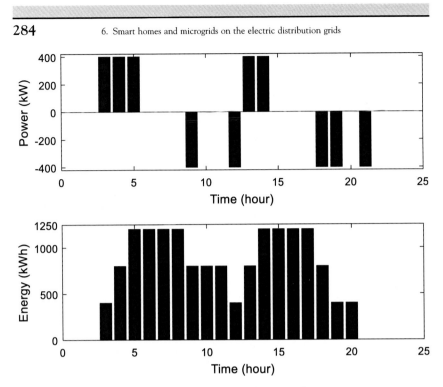

FIGURE 6.12   The operating pattern of battery in microgrid 1.

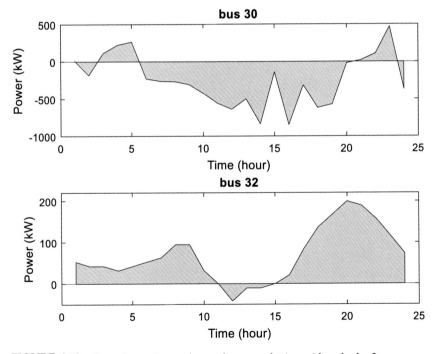

FIGURE 6.13   Operating patterns of smart homes and microgrid on feeder 2.

and it is not economical to buy electricity from the transmission network. The capacity of the battery is 1200 kWh and the battery is charged to the maximum capacity for shifting energy as much as possible.

6.6 Normal operation of feeder 2

In feeder 2, one microgrid and one smart home are installed on buses 30 and 32, respectively. Fig. 6.13 demonstrates the operating patterns of

TABLE 6.7 Produced-consumed powers by smart homes and microgrid on feeder 2.

Hour	Microgrid on bus 30 (kW)	Smart home on bus 32 (kW)
1	10	52.5
2	−184	42
3	112	42
4	222	31.5
5	264	42
6	−230	52.5
7	−264	63
8	−268	94.5
9	−306	94.5
10	−426	31.5
11	−560	0
12	−640	−42
13	−502	−10.5
14	−836	−10.5
15	−140	0
16	−848	21
17	−316	84
18	−620	136.5
19	−572	168
20	−20	199.5
21	20	189
22	110	157.5
23	470	115.5
24	−370	73.5

the smart homes and the microgrid on feeder 2. The microgrid on bus 30 produces power for many hours of the day from hours 5 to 20. The smart home is able to produce power in hours 12–15. In the smart home, it is not possible to shift energy, and all the produced energy must be consumed or sent to the grid. In the microgrid, the energy management system manages and shifts energy based on economic or

TABLE 6.8   Flowed power per kW in lines of feeder 2.

Hour	Line number						
	26	27	28	29	30	31	32
1	175	160	145	115	105	67.5	15
2	−52	−64	−76	−100	84	54	12
3	244	232	220	196	84	54	12
4	321	312	303	285	63	40.5	9
5	396	384	372	348	84	54	12
6	−65	−80	−95	−125	105	67.5	15
7	−21	−45	−69	−117	147	87	24
8	96.5	60.5	24.5	−47.5	220.5	130.5	36
9	126	81	36	−54	252	139.5	45
10	−57	−102	−147	−237	189	76.5	45
11	−200	−248	−296	−392	168	48	48
12	−322	−370	−418	−514	126	6	48
13	−130	−181	−232	−334	168	40.5	51
14	−486.5	−534.5	−582.5	−678.5	157.5	37.5	48
15	175	133	91	7	147	42	42
16	−557	−593	−629	−701	147	57	36
17	83	41	−1	−85	231	126	42
18	−123.5	−171.5	−219.5	−315.5	304.5	184.5	48
19	1	−53	−107	−215	357	222	54
20	629.5	569.5	509.5	389.5	409.5	259.5	60
21	614	560	506	398	378	243	54
22	605	560	515	425	315	202.5	45
23	833	800	767	701	231	148.5	33
24	−139	−160	−181	−223	147	94.5	21

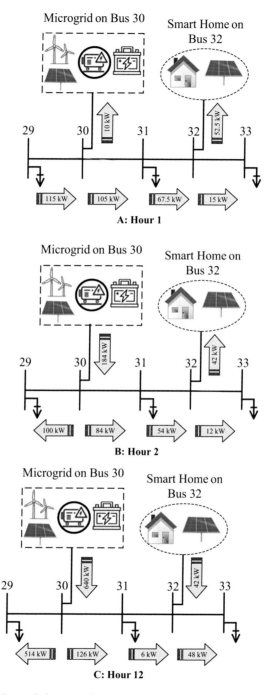

FIGURE 6.14 Power balance in the locations of microgrid on bus 30 and smart home on bus 32.

technical dictates. The numerical data of the produced-consumed powers by the smart homes and the microgrid are listed in Table 6.7.

Feeder 2 comprises 7 lines including lines 26–32 and the flowed power through those lines is presented in Table 6.8. The microgrid is installed on bus 30 which is located between lines 29 and 30. In many hours of the day, the microgrid sends power back to the grid through line 29. The connection between feeder 2 and the upstream grid is made through line 26. The negative powers in line 26 indicate sending power from feeder 2 to the distribution grid. Such power may be consumed by the loads that are located before feeder 2 or may be traded with the transmission network.

Fig. 6.14 demonstrates the power balance in the locations of the microgrid on bus 30 and the smart home on bus 32. At hour 1, both the microgrid and the smart home take power from the grid. At hour 2, the microgrid supplies both the grid and the smart home. The loads after bus 30 including the smart home on bus 32 are supplied by the microgrid. As well, 100 kW is injected into the distribution grid that could be consumed by loads on feeder 2 or by loads on the distribution grid or even may be traded with the upstream transmission network. At hour 12, both the microgrid and smart home produce power and

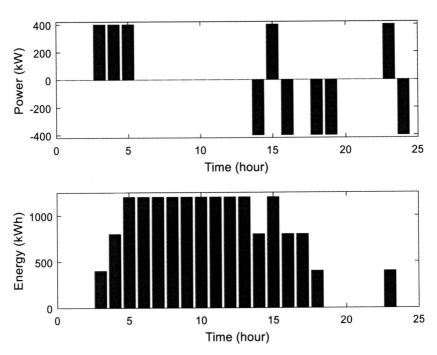

FIGURE 6.15   Charging scheduling and energy of battery in microgrid 2.

supply the loads on feeder 2. The produced power by the smart home is only sufficient to supply the adjacent loads on bus 33. However, the produced power by the microgrid is used to supply many loads on feeder 2.

Fig. 6.15 indicates the charging scheduling and energy of the battery in microgrid 2. It is clear that the battery stores energy at hours 3–5 and uses this energy when electricity is expensive like hours 15–20. The maximum capacity of the battery which is 1200 kWh is used by the energy management system for storing and shifting energy.

TABLE 6.9 Daily energy cost under various operating conditions.

Operating condition	Daily energy cost ($/day)
Nominal operating condition	3493.015
Off-grid operation of feeder 1	3608.520
Off-grid operation of feeder 2	3618.215
Off-grid operation of feeders 1 and 2	3733.720

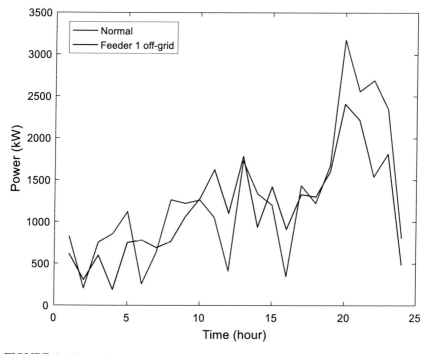

FIGURE 6.16 Exchanged power between the distribution grid and upstream transmission network.

6.7 Off-grid operation of feeder 1

The developed distribution grid is integrated with two feeders and each feeder is able to operate off-grid. In the off-grid operation of feeder 1, the loads on the feeder are supplied by the smart homes and the microgrid. Table 6.9 presents the daily energy cost under various operating conditions. When feeder 1 operates off-grid, the energy cost increases by 3.2% because the off-grid operation utilizes the diesel generator in microgrid 1. The off-grid operation of feeder 2 raises the energy cost by 3.3% and the off-grid operation of both feeders increases the energy cost by 6.8%.

Fig. 6.16 shows the exchanged power between the distribution grid and the upstream transmission network. It is clear that the exchanged power is positive at all hours which means receiving power from the transmission network. During the off-grid operation of feeder 1, the received power from the transmission network reduces because the loads on feeder 1 are not supplied by the transmission network anymore.

Fig. 6.17 shows the produced-consumed powers by microgrid 1 on bus 13 under two cases of normal and off-grid conditions. In the off-grid

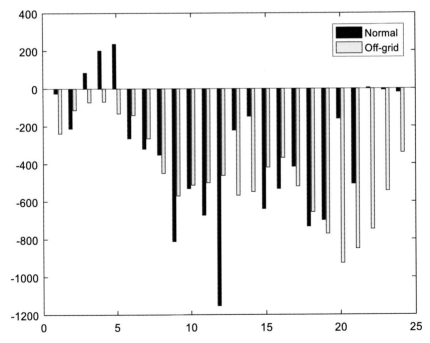

FIGURE 6.17   Produced-consumed powers by microgrid 1 on bus 13.

condition, the microgrid produces power at all hours, while in the normal condition, it takes power from the distribution grid at some hours. In normal conditions, the primary duty of the microgrid is to shift energy from low-priced hours to high-priced hours because the distribution grid is connected to the feeder and there is not any concern regarding the required energy of the loads on the feeder. Consequently, the microgrid properly shifts energy to hours like 15−20 when electricity is

TABLE 6.10   Produced power by microgrid 1 under off-grid operation of feeder 1.

Hour	Microgrid on bus 13 (kW)
1	−237.75
2	−113.4
3	−71.8
4	−69.05
5	−132.6
6	−141.75
7	−265.4
8	−449.3
9	−569.85
10	−512.85
11	−500
12	−462
13	−566.95
14	−548.1
15	−417.5
16	−366.8
17	−519.1
18	−655.5
19	−769.7
20	−925.5
21	−849.5
22	−745.25
23	−542.25
24	−339.25

expensive. On the other hand, in the off-grid operation, the microgrid is utilized for minimizing energy costs as well as supplying all loads on the feeder without any load curtailment. Therefore, the flexibility of the energy management system for shifting energy from low-priced hours to high-priced hours is reduced. It has to shift energy to all 24 hours for supplying the loads. As a result, the energy is shifted to hours like 20–24 for supplying the loads and not for reducing energy costs. Table 6.10 presents the numerical data related to the produced power of microgrid 1 under the off-grid operation of feeder 1.

In the off-grid operation of feeder 1, the diesel generator in microgrid 1 is operated to handle the load demands. The daily operating regime of the diesel generator is shown in Fig. 6.18. It is clear that the diesel generator is operated during on-peak loading when the battery and renewable resources are not sufficient to supply the loads. The diesel generator is operated with maximum capacity at hours 20–23 when the solar energy is zero. The daily operation of the battery is depicted in Fig. 6.19. The battery similar to the diesel generator produces energy at hours 18–23 when renewable energies are at the minimum level as well as the loads are in the maximum demand. The energy stored in the

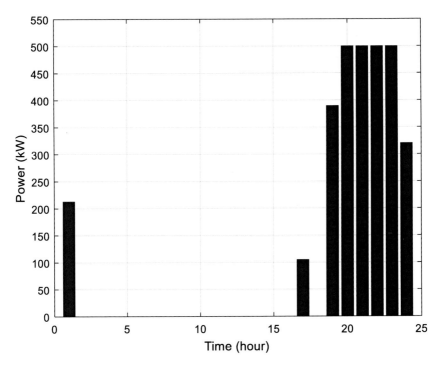

FIGURE 6.18 Daily operating regime of diesel generator in microgrid 1 under off-grid operation of feeder 1.

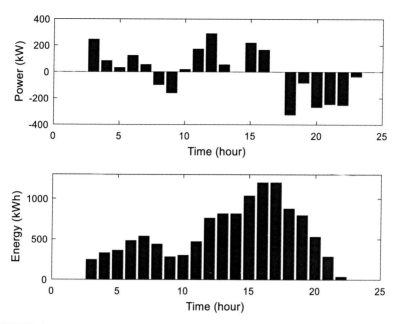

FIGURE 6.19 Daily operating regime of battery in microgrid 1 under off-grid operation of feeder 1.

TABLE 6.11 Flowed power per kW in lines of feeder 1 under off-grid operation.

Hour	6	7	8	9	10	11	12	13	14	15	16	17
						Line number						
1	0	−50	−90	−105	−114	−125.25	−140.25	97.5	67.5	52.5	37.5	22.5
2	0	−40	−24	−36	−14.4	−23.4	−35.4	78	54	42	30	18
3	0	−40	2	−10	27.2	18.2	6.2	78	54	42	30	18
4	0	−30	−8	−17	5.2	−1.55	−10.55	58.5	40.5	31.5	22.5	13.5
5	0	−40	−36	−48	−33.6	−42.6	−54.6	78	54	42	30	18
6	0	−50	−30	−45	−18	−29.25	−44.25	97.5	67.5	52.5	37.5	22.5
7	0	−80	−74	−98	−76.4	−94.4	−118.4	147	99	75	51	27
8	0	−120	−143	−179	−165.8	−192.8	−228.8	220.5	148.5	112.5	76.5	40.5
9	0	−150	−186	−231	−225.6	−259.35	−304.35	265.5	175.5	130.5	85.5	40.5
10	0	−150	−156	−201	−195.6	−229.35	−274.35	238.5	148.5	103.5	58.5	13.5
11	0	−160	−140	−188	−176	−212	−260	240	144	96	48	0
12	0	−160	−120	−168	−156	−192	−240	222	126	78	30	−18

(Continued)

TABLE 6.11   (Continued)

Hour	6	7	8	9	10	11	12	13	14	15	16	17
						Line number						
13	0	−170	−172	−223	−227.2	−265.45	−316.45	250.5	148.5	97.5	46.5	−4.5
14	0	−160	−171	−219	−228.6	−264.6	−312.6	235.5	139.5	91.5	43.5	−4.5
15	0	−140	−110	−152	−134	−165.5	−207.5	210	126	84	42	0
16	0	−120	−98	−134	−114.8	−141.8	−177.8	189	117	81	45	9
17	0	−140	−166	−208	−199.6	−231.1	−273.1	246	162	120	78	36
18	0	−160	−225	−273	−273	−309	−357	298.5	202.5	154.5	106.5	58.5
19	0	−180	−272	−326	−333.2	−373.7	−427.7	342	234	180	126	72
20	0	−200	−345	−405	−435	−480	−540	385.5	265.5	205.5	145.5	85.5
21	0	−180	−320	−374	−404	−444.5	−498.5	351	243	189	135	81
22	0	−150	−290	−335	−374	−407.75	−452.75	292.5	202.5	157.5	112.5	67.5
23	0	−110	−210	−243	−270	−294.75	−327.75	214.5	148.5	115.5	82.5	49.5
24	0	−70	−130	−151	−166	−181.75	−202.75	136.5	94.5	73.5	52.5	31.5

battery reaches the maximum capacity which is 1200 kWh at hours 16−17.

Table 6.11 shows the flowed power in lines of feeder 1 under off-grid operation. The flowed power through line 6 is zero in all hours which means an off-grid condition for feeder 1. The microgrid is located between lines 12 and 13. It is seen that the flowed powers through lines 12 to 6 are negative in all hours which indicates injecting power from the microgrid toward these lines (i.e., from bus 13 toward bus 7). The flowed power through lines 13 to 17 is positive in all hours as well. This point shows that the microgrid is supplying the loads on these lines at all hours. It is seen that the microgrid plays a key role in the off-grid operation of the feeder. The energy management system properly schedules the operating pattern of the battery and diescl generator to deal with loads on all buses of feeder 1 during off-grid operation.

The power balance in the location of the microgrid on bus 13 at hours 4 and 20 is depicted in Fig. 6.20. These two hours are selected because the minimum load demand is seen at hour 4 and the maximum loading condition happens at hour 20. At both hours, the microgrid produces power for supplying the loads in both directions toward the first bus (bus 7) and the last bus (bus 18) of feeder 1. The produced power at hour 20 is significantly more than the power at hour 4. The microgrid needs to employ all its resources to supply such enormous power at hour 20.

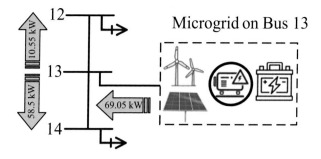

A: Hour 4

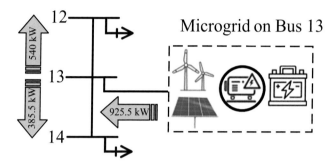

B: Hour 20

FIGURE 6.20   Power balance in location of microgrid on bus 13 under off-grid operation of feeder 1.

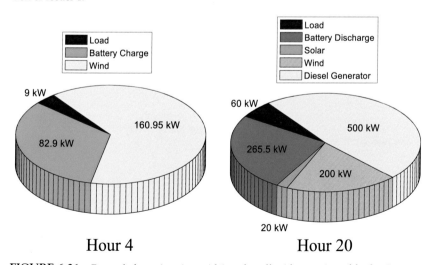

FIGURE 6.21   Power balance in microgrid 1 under off-grid operation of feeder 1.

In order to demonstrate this point, the power balance in microgrid 1 at hours 4 and 20 is demonstrated in Fig. 6.21. At hour 4, the wind energy supplies the battery charging power and the load power. The surplus of power injected to the feeder is 69.05 kW. At hour 20, all resources including the wind, solar, battery, and diesel generator are producing energy. About 60 kW of energy is consumed by loads in the microgrid and the rest of the energy which is 925.5 kW is injected to the feeder.

6.8 Off-grid operation of feeder 2

In the off-grid operation of feeder 2, the loads on the feeder should be supplied by smart home on bus 32 and microgrid on bus 30. Fig. 6.22 demonstrates the exchanged power between the distribution grid and the upstream transmission network. When the feeders become off-grid, the received power from the transmission network reduces substantially because the loads on the feeders are not supplied by the transmission network but are consumed by local resources on the feeders. Additionally, when both feeders are off-grid, there is not any resource on the rest of the

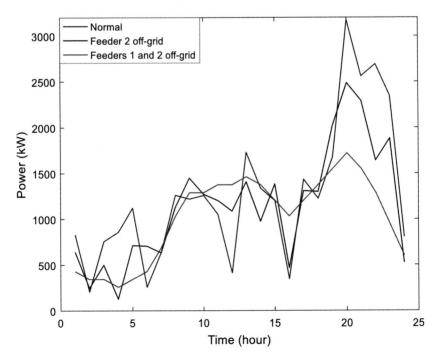

FIGURE 6.22 Exchanged power with upstream transmission network under various operating conditions.

distribution grid to shift or regulate energy. The distribution grid has to receive power from the transmission network directly without any regulation or management. As a result, the power is not shifted from the low-priced hours such as 2–6 to the high-priced hours like 18–22.

Table 6.12 shows the produced power by microgrid 2 under the off-grid operation of feeder 2. In this condition, the microgrid should supply all loads on the feeder and produces power at all hours. Fig. 6.23

TABLE 6.12 Produced power by microgrid 2 under off-grid operation of feeder 2.

Hour	Microgrid on bus 30 (kW)
1	−180
2	−144
3	−144
4	−108
5	−144
6	−180
7	−267
8	−400.5
9	−477
10	−414
11	−408
12	−366
13	−423
14	−397.5
15	−357
16	−327
17	−441
18	−544.5
19	−627
20	−709.5
21	−648
22	−540
23	−396
24	−252

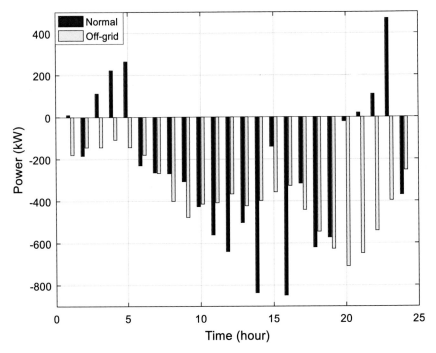

FIGURE 6.23 Produced-consumed power by microgrid 2 under normal and off-grid conditions.

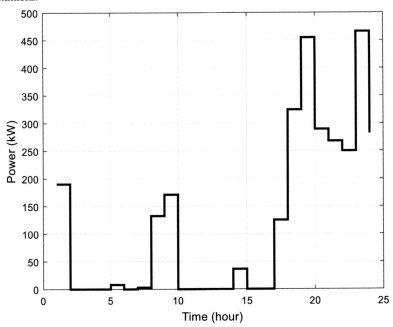

FIGURE 6.24 Daily operating regime of diesel generator in microgrid 2 under off-grid operation of feeder 2.

compares the produced-consumed power by microgrid 2 under normal and off-grid conditions. In the off-grid condition, the microgrid produces power at all hours but in the normal condition, it works on both production and consumption states. In normal conditions, the microgrid takes power from the grid when energy is cheap like hours 3−5 and 22−23. This taken energy is stored and used when energy is expensive such as hours 16−20. On the other hand, the off-grid feeder cannot take any power from the grid and has to supply all the loads by the local microgrid and smart home. Hence, the microgrid cannot take power from the feeder and it only injects power into the feeder.

In the off-grid condition, the microgrid utilizes local resources including the battery and the diesel generator to supply the loads on the microgrid as well as on the feeder. Figs. 6.24 and 6.25 show the 24-hour operating regime for the diesel generator and the battery. The diesel generator is mostly operated at hours 18−24 when the solar energy is on the minimum level or it is zero and on the other hand, the loads are on the maximum level. The battery follows a similar pattern to the diesel generator and discharges power at hours 18−24. The battery charges power when wind and solar energy are at the maximum level and the loads do not consume too much energy.

Table 6.13 presents the flowed power through the lines of feeder 2 under off-grid operating conditions. The flowed power through line 25

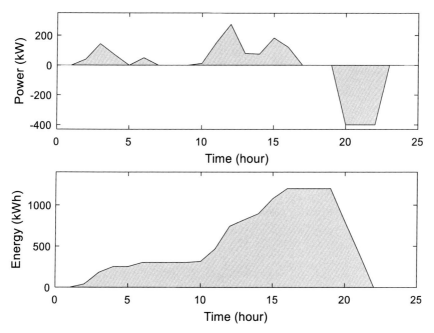

FIGURE 6.25 Daily operating regime of battery in microgrid 2 under off-grid operation of feeder 2.

is zero at all hours which means an off-grid condition for feeder 2. Microgrid 2 is placed on bus 30 between lines 29 and 30. The flowed power through line 29 is negative in all hours which indicates injecting power from the microgrid on bus 30 toward the buses before the microgrid location including buses 29–26. The flowed power through line 30 is positive in all hours which specifies injecting power from the

TABLE 6.13   Flowed power per kW in lines of feeder 2 under off-grid operation.

	Line number							
Hour	25	26	27	28	29	30	31	32
1	0	−15	−30	−45	−75	105	67.5	15
2	0	−12	−24	−36	−60	84	54	12
3	0	−12	−24	−36	−60	84	54	12
4	0	−9	−18	−27	−45	63	40.5	9
5	0	−12	−24	−36	−60	84	54	12
6	0	−15	−30	−45	−75	105	67.5	15
7	0	−24	−48	−72	−120	147	87	24
8	0	−36	−72	−108	−180	220.5	130.5	36
9	0	−45	−90	−135	−225	252	139.5	45
10	0	−45	−90	−135	−225	189	76.5	45
11	0	−48	−96	−144	−240	168	48	48
12	0	−48	−96	−144	−240	126	6	48
13	0	−51	−102	−153	−255	168	40.5	51
14	0	−48	−96	−144	−240	157.5	37.5	48
15	0	−42	−84	−126	−210	147	42	42
16	0	−36	−72	−108	−180	147	57	36
17	0	−42	−84	−126	−210	231	126	42
18	0	−48	−96	−144	−240	304.5	184.5	48
19	0	−54	−108	−162	−270	357	222	54
20	0	−60	−120	−180	−300	409.5	259.5	60
21	0	−54	−108	−162	−270	378	243	54
22	0	−45	−90	−135	−225	315	202.5	45
23	0	−33	−66	−99	−165	231	148.5	33
24	0	−21	−42	−63	−105	147	94.5	21

microgrid on bus 30 toward the buses after the microgrid location including buses 31−33.

Fig. 6.26 shows the power balance in the location of the microgrid on bus 30 under the off-grid operation of feeder 2. It is seen that the microgrid produces energy at both hours 4 and 20, where, hour 4 is the minimum loading condition and hour 20 shows on-peak loading demand. The operation of the microgrid during these two hours is depicted in Fig. 6.27. At hour 4, only the wind generating system is working and it is enough to supply the local loads inside the microgrid including battery charging power and load power. The surplus of wind energy is injected into the feeder which is 108 kW. At hour 20, all resources of the

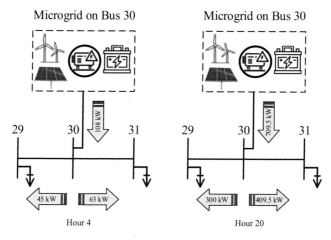

FIGURE 6.26 Power balance in the location of microgrid 2 on bus 30 under the off-grid operation of feeder 2.

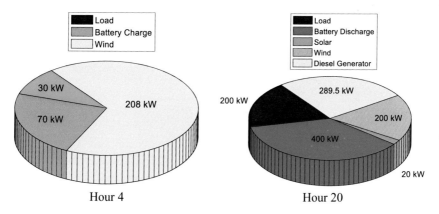

FIGURE 6.27 Power balance in microgrid 2 under the off-grid operation of feeder 2.

microgrid are utilized to supply the loads inside the microgrid as well as outside the microgrid on the feeder. The produced powers by the battery and the diesel generator are controlled to compensate for and balance the shortage of wind-solar energy.

6.9 Impacts of battery on the system

In microgrids, the battery plays a key role in shifting and managing energy. In order to assess the impacts of the battery on the energy

TABLE 6.14   Daily energy cost under various battery capacities.

Capacity of battery in both microgrids	Daily energy cost ($/day)
1200 kWh	3493.015
1800 kWh	3373.015
2400 kWh	3253.015

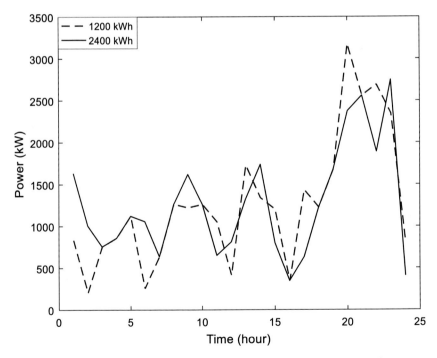

FIGURE 6.28   Exchanged power with upstream transmission network under various battery capacities.

management system, the energy cost under various battery capacities is presented in Table 6.14. With a 1200 kWh battery on both microgrids, the energy cost is 3493.015 $/day. Increasing battery capacity by 50% decreases the energy cost to 3373.015 $/day which shows a 3.4% reduction and escalating battery capacity by 100% cuts the energy cost by 6.9%. The larger battery allows the energy management system for storing and shifting more energy resulting in less energy cost.

Fig. 6.28 represents the power from the upstream transmission network to the distribution grid with two different battery capacities in the microgrids. The larger battery obviously shifts higher levels of energy and changes the pattern of power between the transmission network and the distribution grid. During the time periods when the energy is

TABLE 6.15 Energy cost for off-grid operation of feeder 1 with various battery capacities in microgrid 1.

Battery capacity in microgrid 1	Daily energy cost ($/day)
1200 kWh	3608.520
1800 kWh	3586.957

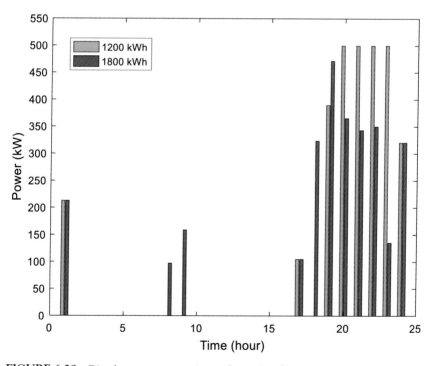

FIGURE 6.29 Diesel generator operation under various battery capacities.

expensive such as hours 16–22, the larger battery can efficiently regulate the received power from the transmission network.

Under the off-grid operation of feeder 1, the battery and diesel generator show complementary operating patterns. Once the battery is not able to handle the loads through the discharging operation, the diesel generator complements the battery operation. Accordingly, the smaller battery capacity results in more operation for the diesel generator and vice-versa. Table 6.15 indicates the energy cost under various battery capacities in microgrid 1 when feeder 1 is operated off-grid. As it was stated, the larger battery reduces the energy cost by reducing diesel generator operation. The diesel generator operation under two different battery capacities is depicted in Fig. 6.29. It is clear that the microgrid with a larger battery takes less power from the diesel generator. The maximum capacity of the diesel generator which is 500 kW is not reached when the microgrid is equipped with an 1800 kWh battery and this point demonstrates that the microgrid may be equipped with a smaller size diesel generator. On the other hand, when the microgrid is equipped with a 1200 kWh battery, the maximum capacity of the diesel generator is used for many hours.

Figs. 6.30 and 6.31 show the battery operation under two capacities of 1200 and 1800 kWh. Both batteries show a similar pattern and shift

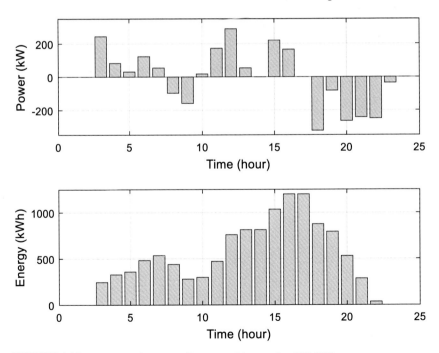

FIGURE 6.30   Power and energy of battery with capacity 1200 kWh.

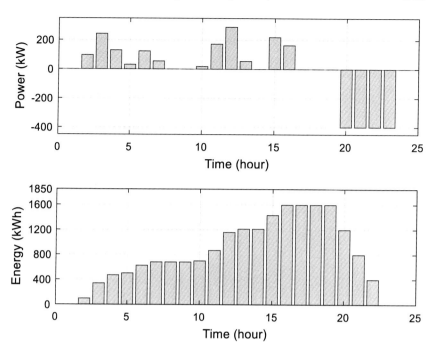

FIGURE 6.31   Power and energy of battery with capacity 1800 kWh.

TABLE 6.16   Impact of battery in microgrids 1 and 2 on energy cost.

Operating condition	Daily energy cost ($/day)
One 1200 kWh battery in both microgrids	3493.015
No battery in both microgrids	3733.015

energy to the last hours of the day. The battery with a smaller capacity does not have enough capacity to shift large amounts of energy over the day hours and it has to charge and discharge energy in several hours. The battery with larger capacity stores energy during day hours and discharges this energy at the last hours.

The distribution grid without the battery in the microgrids is not able to shift and manage energy. This point makes a negative impact on the energy cost of the distribution grid. Table 6.16 presents the energy cost of the distribution grid with and without the battery in the microgrids. Removing the battery from microgrids increases the daily energy cost of the distribution grid by 6.4%. Fig. 6.32 shows the exchanged power between the transmission network and the distribution grid with and without the battery in microgrids. When the microgrids are not

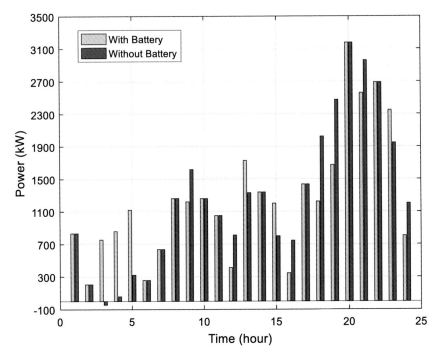

FIGURE 6.32   Exchanged power with upstream transmission network with and without the battery in microgrids.

equipped with a battery, the energy cannot be sifted over the 24-hour. The battery takes energy from the transmission network when it is inexpensive such as hours 3–5 but the system without the battery cannot use this opportunity for consuming such cheap energy.

6.10 Stochastic programming

In stochastic programming, some parameters are associated with uncertainty. The load, wind, and solar powers are modeled as uncertain parameters and defined by a set of scenarios of performance. The mathematical model of deterministic programming needs to be reformulated to achieve stochastic programming. The cost of purchasing electricity by distribution grid is reformulated as Eq. (6.20) which shows the expected value of cost instead of the true value of cost which was already calculated in Eq. (6.1). The relationships in Eqs. (6.4) to (6.11) are integrated with index "s" which is the index of scenarios and those relationships are defined over the set of scenarios. The relationships related to the battery and the diesel generator are already expressed through Eqs. (6.12) to (6.19) are not changed and they are not integrated with index "s." In other words, the battery and the diesel generator are not affected by uncertainty in the parameters and they do not change their operations under scenarios of performance.

$$z_{ele} = \sum_{s=1}^{S} \left[\left\{ \sum_{t=1}^{T} \left(P_f^{1,2,t,s} \times C_{ele}^t \times k_{pr} \right) \right\} \times O_{pr}^s \right] \tag{6.20}$$

Similar to the previous chapters, seven scenarios of performance are defined for modeling the uncertainty created by load, wind, and solar powers. Fig. 6.33 shows the probability of occurrence for each scenario of performance. Fig. 6.34 indicates the scenarios of performance that are related to load, solar, and wind powers, where the black solid line shows the main scenario of performance which is the deterministic operating condition.

Table 6.17 presents the numerical values of load power under each scenario of performance. The data of solar and wind powers under each scenario of performance are listed in Tables 6.18 and 6.19. The first scenario is regarded as the main scenario of performance which is the deterministic operating condition. The other scenarios are achieved by changing the data of the first scenario around the nominal values by generating random numbers from the standard normal distribution.

The distribution grid with and without uncertainty in the parameters is simulated and compared in Table 6.20. The stochastic programming

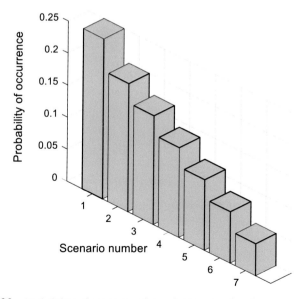

FIGURE 6.33 Probability of occurrence for each scenario of performance.

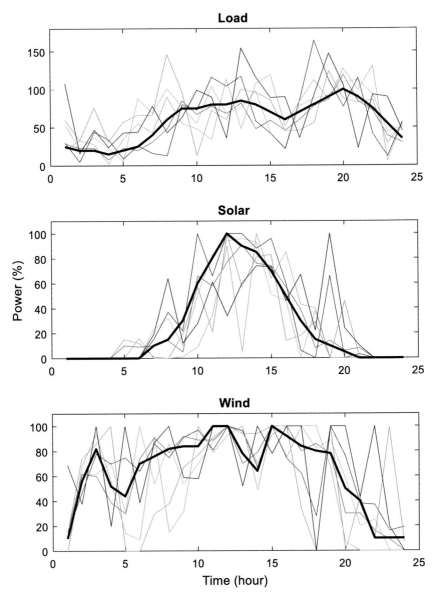

FIGURE 6.34 Scenarios of performance related to load, solar, and wind powers. *Black solid line*: Main scenario of performance showing deterministic operating condition.

which comprises uncertainty in the parameters results in more energy cost compared to the model without uncertainty. The uncertainty increases the daily energy cost by about 240 $/day or 6.8%. The distribution grid with uncertainty in the parameters needs further utilization

TABLE 6.17   Load power in percentage under each scenario.

Hour	Scenario number						
	1	2	3	4	5	6	7
1	25	29	55	30	60	21	108
2	20	23	21	5	33	33	16
3	20	23	37	44	76	28	47
4	15	8	23	34	31	2	24
5	20	23	6	9	57	35	43
6	25	20	88	26	66	55	44
7	40	31	69	16	66	43	78
8	60	48	101	13	146	90	43
9	75	85	77	80	97	55	63
10	75	67	55	73	13	48	99
11	80	78	104	116	72	34	90
12	80	68	68	104	60	113	36
13	85	82	76	117	100	48	154
14	80	48	98	58	96	107	116
15	70	59	42	42	86	119	89
16	60	46	21	22	45	52	90
17	70	60	78	85	46	84	36
18	80	78	88	164	56	112	93
19	90	87	92	112	124	86	147
20	100	119	110	77	88	127	106
21	90	84	112	115	82	99	23
22	75	72	94	41	86	131	93
23	55	39	44	12	29	6	90
24	35	30	51	57	54	55	45

of local resources to handle the uncertainty and this point increases the energy cost. In the deterministic model, the optimal operating condition with minimum cost is achieved for the system and all operating patterns of resources are optimized. When the grid is faced with uncertainty, the stochastic programming has to change the operating patterns

TABLE 6.18 Solar power in percentage under each scenario.

Hour	Scenario number						
	1	2	3	4	5	6	7
1	0	0	0	0	0	0	0
2	0	0	0	0	0	0	0
3	0	0	0	0	0	0	0
4	0	0	0	0	0	0	0
5	0	10	0	0	15	0	0
6	0	9	0	0	12	16	0
7	10	10	23	15	1	0	19
8	15	14	32	37	0	0	64
9	30	5	10	22	19	1	12
10	60	66	51	100	57	9	28
11	80	58	83	66	21	35	61
12	100	77	94	100	100	90	34
13	90	91	96	100	57	5	60
14	85	76	93	88	81	100	74
15	70	74	51	96	83	64	73
16	50	57	46	60	14	85	46
17	30	39	0	6	18	43	67
18	15	8	38	0	29	40	22
19	10	14	0	66	0	0	100
20	5	6	0	0	46	0	24
21	0	8	0	3	1	0	10
22	0	0	0	0	0	0	0
23	0	0	0	0	0	0	0
24	0	0	0	0	0	0	0

of resources from the optimal condition to a new nonoptimal point in order to handle the uncertainty. Such alteration in the operating patterns of resources increases the energy cost compared to the deterministic model.

Fig. 6.35 shows the exchanged power between the transmission network and the distribution grid under all scenarios of performance.

TABLE 6.19   Wind power in percentage under each scenario.

Hour	Scenario number						
	1	2	3	4	5	6	7
1	10	16	0	15	15	0	69
2	56	62	61	47	74	68	38
3	82	60	62	78	86	94	100
4	52	39	71	70	100	56	20
5	44	30	31	75	0	14	100
6	70	74	100	64	0	9	39
7	76	91	54	86	30	92	87
8	82	79	86	75	38	15	100
9	84	92	54	91	67	32	59
10	84	89	70	97	77	100	58
11	100	89	88	81	91	79	100
12	100	100	100	100	100	93	100
13	78	70	97	96	94	100	52
14	64	79	70	67	94	49	100
15	100	100	93	100	100	100	78
16	92	100	100	74	100	75	100
17	84	80	100	62	33	100	100
18	80	86	82	28	0	100	0
19	78	77	89	100	100	67	100
20	50	30	58	76	0	7	100
21	40	30	20	38	43	0	43
22	10	16	25	37	11	0	100
23	10	9	0	16	99	0	19
24	10	0	0	19	7	0	0

The black solid line shows the main scenario of performance related to the deterministic operating condition. Stochastic programming varies the exchanged power from one scenario to another one to deal with the uncertainty. In some scenarios of performance, the power is significantly different from the deterministic operating condition.

TABLE 6.20 Operation of the system with and without uncertainty in parameters.

Model	Daily energy cost ($/day)
Deterministic programming	3493.015
Stochastic programming	3733.305

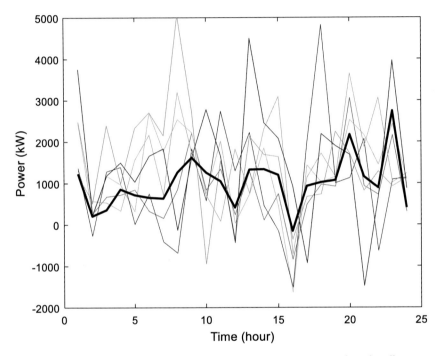

FIGURE 6.35 Exchanged power with upstream transmission network under all scenarios of performance. *Black solid line*: Main scenario of performance showing deterministic operating condition.

Figs. 6.36 and 6.37 demonstrate the produced-consumed powers by microgrids 1 and 2 under all scenarios of performance. It is seen that the exchanged power between the microgrids and the distribution grid changes significantly under various scenarios of performance. The variations and uncertainties in the loads and renewable energies are dealt with by such alterations of exchanged power. As it was already stated, the diesel generator and battery operations are not affected by uncertainty in the parameters and they show a unique and single operating pattern under all scenarios of performance. As a result, the uncertainty needs to be dealt with by the exchanged power between the microgrids and the distribution grid as is seen in the results.

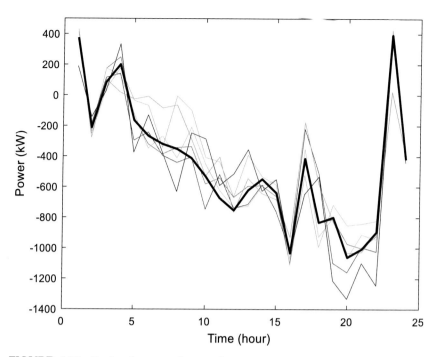

FIGURE 6.36 Produced-consumed power by microgrid 1 under all scenarios of performance. *Black solid line*: Main scenario of performance showing deterministic operating condition.

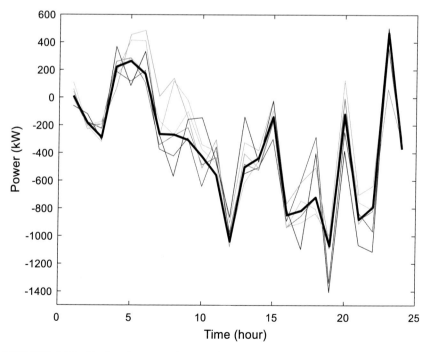

FIGURE 6.37 Produced-consumed power by microgrid 2 under all scenarios of performance. *Black solid line:* Main scenario of performance showing deterministic operating condition.

Tables 6.21 and 6.22 present the operation of the battery and the diesel generator in the microgrids under uncertainty in the parameters. It is clear that the battery and the diesel generator show a unique operating pattern under all uncertainties in the parameters which are modeled by scenarios of performance. The battery charges energy in the initial hours of 24-hour and discharges this energy at hours 20–22 when it is not economical to buy electricity from the upstream grid. The maximum

TABLE 6.21   Operation of resources in microgrid 1 under uncertainty in parameters.

Hour	Diesel generator (kW)	Battery charging power (kW)	Battery discharging power (kW)	Battery energy (kWh)
1	0	400	0	400
2	0	0	0	400
3	0	400	0	800
4	0	400	0	1200
5	0	0	0	1200
6	0	0	0	1200
7	0	0	0	1200
8	0	0	0	1200
9	0	0	0	1200
10	0	0	0	1200
11	0	0	0	1200
12	0	0	0	1200
13	0	0	0	1200
14	0	0	0	1200
15	0	0	0	1200
16	500	0	0	1200
17	0	0	0	1200
18	500	0	0	1200
19	500	0	0	1200
20	500	0	400	800
21	500	0	400	400
22	500	0	400	0
23	0	400	0	400
24	0	0	400	0

TABLE 6.22 Operation of resources in microgrid 2 under uncertainty in parameters.

Hour	Diesel generator (kW)	Battery charging power (kW)	Battery discharging power (kW)	Battery energy (kWh)
1	0	0	0	0
2	0	0	0	0
3	0	0	0	0
4	0	400	0	400
5	0	400	0	800
6	0	400	0	1200
7	0	0	0	1200
8	0	0	0	1200
9	0	0	0	1200
10	0	0	0	1200
11	0	0	0	1200
12	0	0	400	800
13	0	0	0	800
14	0	0	0	800
15	0	400	0	1200
16	0	0	400	800
17	500	0	0	800
18	500	0	0	800
19	500	0	400	400
20	500	400	0	800
21	500	0	400	400
22	500	0	400	0
23	0	400	0	400
24	0	0	400	0

capacity of the battery which is 1200 kWh is used by the microgrid. Both microgrids utilize the diesel generator at hours 16–22 when the loads are on the maximum level. Comparing the results to the deterministic model verifies that the stochastic model needs to utilize the diesel generator to deal with load-energy uncertainty but the deterministic model does not run the diesel generator.

6.11 Conclusions

In this chapter, the integration of microgrids and smart homes into the electric distribution grid was addressed. The 33-bus active distribution grid was considered as a case study and it was formed with two feeders. The first feeder was integrated with 3 smart homes on buses 8, 10, and 18 and one microgrid on bus 13. The second feeder was integrated with one microgrid on bus 30 and one smart home on bus 32. Each microgrid was equipped with a diesel generator, wind generating system, solar generating system, and battery energy storage system. The smart home on bus 8 was supplied by both wind and solar generating systems. The smart home on bus 10 was integrated with the wind-generating system. The smart homes on buses 18 and 32 were equipped with a solar-generating system. The feeders were able to continue operation when the upstream transmission network faced a blackout. The energy management system minimized the energy cost of the active distribution grid when feeders 1 and 2 were connected to the grid. During the off-grid operation of feeders, the energy management system utilized the local resources on each feeder for supplying the loads. As well, the operating cost of diesel generators was minimized.

In normal operating conditions, the smart homes and microgrids sent the surplus of energy to the distribution grid. In the smart homes, the surplus of energy could not be shifted or regulated and it was directly injected into the distribution grid because the smart homes were only equipped with renewable resources. The energy management system planned to extract maximum power from those resources. On the other hand, in the microgrids, the operating patterns of the diesel generator and the battery were optimized by the energy management system. As a result, the surplus of energy in the microgrids was regulated and managed and it was injected into the distribution grid when necessary. At some hours, the microgrids injected an adequate level of power to the distribution grid for supplying the loads on the feeder as well as exchanging power with the transmission network. In normal operating conditions, the diesel generator was not operated and the charging scheduling of the battery was optimized for shifting energy from low-priced hours such as 3−5 to high-priced hours like 18−21.

In the off-grid operation of feeders, the loads on the feeders were supplied by smart homes and microgrids. The off-grid operation of feeder 1 increased the energy cost by 3.2% because of the diesel generator operation. The diesel generator was operated with maximum capacity at hours 20−23 when the solar energy was zero. The off-grid operation of feeder 2 raised the energy cost by 3.3%, and the off-grid operation of both feeders at the same time increased the energy cost by

6.8%. In the off-grid condition, the microgrids needed to produce power at all hours while in the normal condition, the microgrids mostly produced power at the hours when electricity was expensive and they took power from the distribution grid when the electricity was low-priced.

Increasing the battery capacity by 50% and 100% decreased the energy cost by 3.4% and 6.9%, respectively. Removing the battery from microgrids increased the energy cost by 6.4%. During off-grid operation, reducing battery capacity caused more operation for the diesel generator and vice-versa.

In stochastic programming, the load, wind, and solar powers were modeled as uncertain parameters. The uncertainty increased the energy cost by 6.8%. The power between the transmission network and the distribution grid changed in the company with the uncertainty in the parameters to deal with such uncertainty. Whereas, the diesel generator and the battery showed unique and single operating patterns under all scenarios of performance.

References

[1] Mehrjerdi H, Hemmati R, Farrokhi E. Nonlinear stochastic modeling for optimal dispatch of distributed energy resources in active distribution grids including reactive power. Simulation Modelling Practice and Theory 2019;94:1−13.
[2] Ding T, Wang Z, Qu M, Wang Z, Shahidehpour M. A sequential black-start restoration model for resilient active distribution networks. IEEE Transactions on Power Systems 2022;37(4):3133−6.
[3] Wang S, Luo F, Dong ZY, Ranzi G. Joint planning of active distribution networks considering renewable power uncertainty. International Journal of Electrical Power & Energy Systems 2019;110:696−704.
[4] Hemmati R, Saboori H. Emergence of hybrid energy storage systems in renewable energy and transport applications−a review. Renewable and Sustainable Energy Reviews 2016;65:11−23.
[5] Saboori H, Hemmati R. Optimal management and planning of storage systems based on particle swarm optimization technique. Journal of Renewable and Sustainable Energy 2016;8(2).
[6] Shaker MH, Farzin H, Mashhour E. Joint planning of electric vehicle battery swapping stations and distribution grid with centralized charging. Journal of Energy Storage 2023;58:106455.
[7] Saboori H, Hemmati R. Maximizing DISCO profit in active distribution networks by optimal planning of energy storage systems and distributed generators. Renewable & Sustainable Energy Reviews 2017;71:365−72.
[8] Mishra DK, Ghadi MJ, Azizivahed A, Li L, Zhang J. A review on resilience studies in active distribution systems. Renewable and Sustainable Energy Reviews 2021;135:110201.
[9] Mehrjerdi H, Mahdavi S, Hemmati R. Resilience maximization through mobile battery storage and diesel DG in integrated electrical and heating networks. Energy. 2021;237:121195.
[10] Hemmati R. Stochastic energy investment in off-grid renewable energy hub for autonomous building. IET Renewable Power Generation 2019;13(12):2232−9.

[11] Hemmati R, Mehrjerdi H, Nosratabadi SM. Resilience-oriented adaptable microgrid formation in integrated electricity-gas system with deployment of multiple energy hubs. Sustainable Cities and Society 2021;71:102946.

[12] Kim H, Choi H, Kang H, An J, Yeom S, Hong T. A systematic review of the smart energy conservation system: from smart homes to sustainable smart cities. Renewable and Sustainable Energy Reviews 2021;140:110755.

[13] Lin C, Wu W, Guo Y. Decentralized robust state estimation of active distribution grids incorporating microgrids based on PMU measurements. IEEE Transactions on Smart Grid 2020;11(1):810−20.

[14] Ghadi MJ, Ghavidel S, Rajabi A, Azizivahed A, Li L, Zhang J. A review on economic and technical operation of active distribution systems. Renewable and Sustainable Energy Reviews 2019;104:38−53.

[15] Hemmati R, Mahdavi S. Hybrid renewable/nonrenewable/storage resources in electrical grid considering active-reactive losses and depth of discharge. International Journal of Energy Research 2021;45(14):20384−99.

Control and stability of residential microgrids

Energy Management in Homes and Residential Microgrids
DOI: https://doi.org/10.1016/B978-0-443-23728-7.00008-4

319

© 2024 Elsevier Inc. All rights reserved.

Nomenclature

Wind system

A Swept area (m^2)
C_p Power coefficient
P_m Generated power by turbine (W)
V_ω Wind speed (m/s)
ρ Density of air (kg/m^2)
ω_r Rotating speed of wind turbine (m/s)
θ Pitch angle (deg)
λ Tip-speed ratio

Solar system

a Coefficient for temperature under short circuit current (A/C^0)
E_g Band-gap energy
ID Diode current (A)
Ipv Solar panel current (A)
K_C Concentration ratio
K_B Boltzmann constant (jk^{-1})
n Ideality coefficient of diode
P_{PV} Solar panel power (W)
q Electron charge (C)
R Solar irradiation (kW/m^2)
R_S Series resistance of solar panel (Ohm)
T Absolute value of temperature (K)
T_{Cref} Reference value of temperature (C^0)
Vpv Solar panel voltage (V)
V_{MPPT} Traced voltage (V)

Synchronous generator

I_{ds} Current in d-axis of stator (A)
Iqs Current in q-axis of stator (A)
I_{WT}^{REF} DC current in output of wind generating system (A)
I_{WT}^* Computed current for activating converter of wind generating system (A)
Ld Inductance related to d-axis of stator (H)
Lq Inductance related to q-axis of stator (H)
P Number of pole pairs
Rs Resistance related to stator (Ohm)
Vds Voltage related to d-axis of stator (V)
Vqs Voltage related to q-axis of stator (V)

V_{WT} DC voltage in output of wind generating system (A)
λm Rotor flux (wb^2)
T_e Electrical torque (N.m)
ωs Angular velocity (rad/s)

Control systems

I_{dc} DC bus current (A)
I_{bat} Battery current (A)
I_i^L Line current through phase i (A)
I_d Current of d-axis (A)
I_i^{pcc} PCC current through phase i (A)
I_d^{ref} Referenced current through d-axis (A)
I_{dc}^{ref} Referenced level for DC current (A)
I_i Current through phase i (A)
I_d^* Finalized current through d-axis (A)
I_i^* Finalized current through phase i (A)
I_{bat}^{ref} Referenced current through battery (A)
I_{bat}^{reg} Regulated current through battery (A)
I_{In} Inverter current (A)
P_{PV} PV power (W)
P_{wt} Wind generating system power (W)
P_{bat} Battery power (W)
P_i^{pcc} PCC active power through phase i (W)
P_d d-axis power (W)
P_{ref} Referenced level for active power (W)
P_L^{ref} Referenced level for load power (W)
P_g Power of utility grid (W)
V_{dc} Voltage level on DC bus (V)
V_{dc}^{ref} Referenced level for DC bus voltage (V)
R Line resistance (Ohm)
X Line reactance (Ohm)

7.1 Introduction

Deploying the distributed generations (DGs) at a variety of scales in the electrical networks is one of the best approaches for producing clean electricity, cutting carbon emissions, and decreasing global warming. Together with population growth, the increasing need for electricity, and the difficulties associated with electricity generation and transmission; microgrids (MGs) are broadly developed to deal with such issues [1]. The MGs are designed for residential or industrial levels. In addition to providing inexpensive electricity at low prices, the MGs are economical and do not need lengthy transmission lines, where, the investment costs associated with new lines are significantly reduced. The MGs are ecologically friendly since they make wide-ranging use of sustainable and renewable energy systems. They also produce and deliver reliable energy to the district loads [2].

The building-integrated MGs are among the promising structures for demand-side management. MG energy management techniques are one of the most interesting subjects in electrical grids. The DC MGs are among the most efficient structures among the MG systems [3]. The use of DC MGs to link DC devices such as photovoltaics (PVs) and energy storages (ESs) can decrease the application of power electronic switches resulting in efficient and cost-effective distribution grids [4].

The optimal sizing of DGs and ESs is necessary for the DC MGs because the electricity generation and loads are unpredictable and often associated with uncertainty. The DC MGs require proper strategies for controlling energy sources in order to achieve optimal performance [5]. One of these strategies is the battery control and management system. In residential buildings, solar thermal panels, heat pumps, and ES batteries are the most common sources. The load demand decrease and supplying the existing load demand by means of renewable power are two major ways of reducing fossil fuel consumption in buildings [6].

Solar cells and wind turbines are the most commonly used DGs in MGs. A maximum power point tracking (MPPT) system can be used to harvest maximum output power from the solar cell. Numerous MPPT methods have been planned such as short-circuit current method, fuzzy logic technique, and perturbation−observation (P&O) method [7]. There are currently two sorts of generators that are used in wind generating systems including permanent magnet synchronous generators (PMSGs) and variable speed wind turbines based on the doubly fed induction generators [8]. As a result of the MPPT control method in the wind generating systems, the mechanical rotation speed is adjusted according to the actual wind speed to increase the power conversion efficiency. In order to enhance the total efficiency of wind turbines, an operative MPPT control scheme is necessary. The maximum power can be extracted from the wind-generating systems using conventional controllers based on linear techniques such as conventional vector control [9].

The primary duty of a battery is to store energy but it is also able to carry out other functions. The proper control system on the battery enables it to be operated in a variety of network applications like voltage and frequency support. The battery not only is appropriate for large-scale applications but also beneficial at the commercial, industrial, and residential levels for adjusting load demand and expanding renewable energy penetration [10].

There are more issues and complexities associated with island MGs in terms of control, utilization, and energy management. Various parameters such as voltage and frequency are controlled by the energy management system in such grids. It has been shown that the abnormalities caused by a failure in the off-grid residential DC MGs reduce the quality of power by dropping the MG voltage under the nominal level. In off-grid MGs, maintaining the equilibrium between power supply and electricity demand is an inevitable point that should be taken into consideration. The voltage often drops in the off-grid systems when the load increases or generation decreases [11]. Table 7.1 summarizes the research studies in the literature on this subject.

7.2 Network structure

The proposed model for the residential MG with wind and solar energy along with battery storage is shown in Fig. 7.1. The MG is designed with two buses, one AC bus, and one DC bus. The DC bus contains the wind and solar DGs as well as the battery, while the AC bus contains the three-phase electrical load and the three-phase network. A three-phase inverter provides the link between the DC and AC buses. The MG is capable of operating in both off-grid and grid-tied modes. The network can either be connected to or disconnected from the bus. In the island mode, the DGs on the DC bus are responsible for supplying the loads on the AC bus through the inverter, while in the grid-connected mode, the loads are supplied from the utility grid side. In this mode of operation, the wind turbine and solar cell only provide power to the battery for storing energy.

7.3 Solar cell model

The PV arrays are formed by multiple solar cells connected in series or parallel to achieve the required high current, voltage, and power levels. The solar cells act as p−n junction diodes that are fabricated with semiconductor materials. As a result of the PV effect, the current is produced in the solar cell by the absorption of light at the connection point. This current is presented in Eq. (7.1), the diode current is expressed by Eq. (7.2), and the output power of the solar cell is specified by Eq. (7.3) [2]. The equivalent circuit model of a PV cell is presented in Fig. 7.2.

TABLE 7.1 Comparison of research studies in the literature.

References	MPPT control on solar cell	MPPT control on wind turbine	Pitch-angle control on wind turbine	Coordinated control in grid-tied	Coordinated control in off-grid	Active power exchange with network	Mutual support between DGs
[1]	Yes	No	No	Yes	Yes	Yes	No
[2]	Yes	Yes	Yes	No	Yes	No	Yes
[3]	Yes	No	No	No	Yes	No	Yes
[4]	Yes	Yes	Yes	No	Yes	No	Yes
[5]	No	No	No	Yes	Yes	No	No
[6]	No	Yes	Yes	No	Yes	No	Yes
[7]	Yes	No	No	No	Yes	No	No
[8]	No	Yes	No	Yes	No	No	No
[9]	No	Yes	No	Yes	No	No	No
[10]	Yes	Yes	Yes	Yes	Yes	No	Yes
[11]	Yes	No	No	Yes	No	No	No
[12]	No	Yes	No	Yes	No	No	No

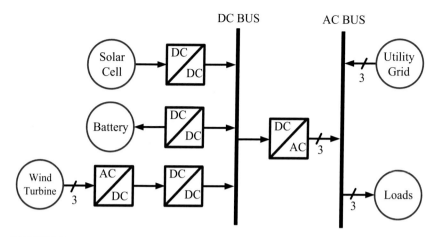

FIGURE 7.1   Outline of microgrid under study.

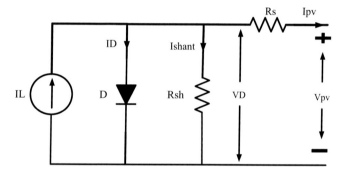

FIGURE 7.2   Simple equivalent circuit model for a photovoltaic cell.

$$Ipv = Rk_c\left[I_{sq} + a(Tc - Tc_{ref})\right] - ID - Ish \tag{7.1}$$

$$ID = \left[K * T^{(3+\gamma/2)}\left[\exp(-\frac{E_g}{nK_BT})\right]\right] \times \left[\exp(\frac{q(Vpv + (Ipv * Rs))}{nK_BT})\right] \tag{7.2}$$

$$Ppv = Vpv * Ipv \tag{7.3}$$

7.3.1 Maximum power point tracking in solar-cell

A P&O method is usually used for MPPT. This technique involves the injection of a small perturbation to cause the PV module power to vary. The PV output power is continuously measured and compared to the determined power at the previous time interval. While the output power rises, the same process continues, otherwise the applied perturbation will

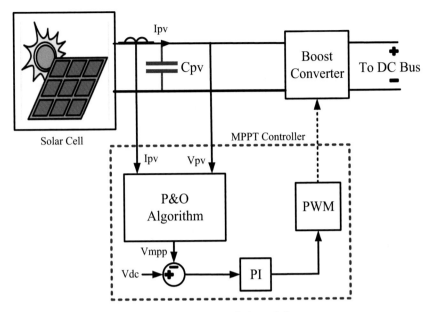

FIGURE 7.3 The designed MPPT for photovoltaic module.

be reversed. The perturbation is applied to the voltage of the PV system. In order to determine whether the power is increasing or decreasing, the PV module voltage is increased or decreased. In the case of an increase in the voltage which causes a rise in the power, this means that the PV module is working on the left side of the maximum power point, which requires additional perturbations to reach the maximum power point. When a voltage rise results in a decrease in the power, the PV module is working on the right side of the maximum power point, and additional perturbations towards the left side are needed to reach the maximum power point [7].

Fig. 7.3 shows the created control system for MPPT in the PV module. In order to acquire a higher DC voltage on the DC bus, the PV system output is linked to the DC–DC converter. A small-size capacitor is installed on the input of the converter for eliminating the high-frequency ripple contents. After comparing the detected voltage to the output voltage of PV according to Eq. (7.4), the error signal passes across a proportional-integral (PI) controller and the proper pulses are generated for the PWM system of the boost converter [1].

$$V_{er} = \left(V_{dc} - V_{mpp}\right)\left(K_p + \frac{K_i}{s}\right) \tag{7.4}$$

7.4 Wind turbine model

The generator plays a key role in the wind-generating system and converts mechanical energy to electrical energy. In the wind generating system, the highly fluctuating torque is applied on the rotor of the generator due to wind speed alterations. As a result, the generator differs significantly from the other generators used in electrical systems. There are many types of generators such as DC, synchronous, induction, permanent magnets, brushless, and synchronous. However, the overall wind-generating system must meet specific performance criteria for connection to the utility grid safely and effectively [12].

The objective of this chapter is to describe a gearless wind energy conversion system that utilizes a PMSG-wind turbine, where the power taken from the wind turbine is injected into the PMSG, and the electrical power produced by PMSG is afterward injected into the DC bus via the boost converter. The relationship presented by Eq. (7.5) expresses the output power of the wind turbine. The general relationships of the wind turbine can be seen through Eqs. (7.6)–(7.8). The mathematical model of PMSG is presented by Eqs. (7.9) and (7.10), and the electromagnetic torque is provided by Eq. (7.11) [2]. The electrical and mechanical data of the wind turbine and generator are listed in Table 7.2.

$$Pm = 0.5 \, \rho \, Cp \, V_\omega^3 \tag{7.5}$$

$$Cp(\lambda, \theta) = 0.73 \left[\frac{151}{\lambda_i} - 0.58\theta - 0.002\theta^{2.14} - 13.2 \right] \exp\left(-\frac{18.4}{\lambda_i} \right) \tag{7.6}$$

$$\frac{1}{\lambda_i} = \frac{1}{\lambda + 0.08\theta} - \frac{0.035}{\theta^3 + 1} \tag{7.7}$$

TABLE 7.2 Wind turbine and permanent magnet synchronous generator parameters.

	Nominal mechanical output power	Base power of electrical generator	Base wind speed	Maximum power at base wind speed	Base rotational speed	Nominal frequency
Wind turbine	8.5 (kW)	8.5/0.9 (kW)	13 (m/s)	0.8 (p.u.)	1 (p.u.)	50 (Hz)
	Stator-phase resistance	Armature inductance	Flux linkage	Phase-phase voltage	Inertia coefficient	Pole pairs
Generator	0.425 (Ohm)	0.835 (mH)	0.433	392 (V)	0.01197 (kg.m²)	5

$$\lambda = \frac{R * \omega_r}{V_\omega} \tag{7.8}$$

$$V_{ds} = R_s i_{ds} + L_d \frac{d}{dt} i_{ds} - \omega_s L_q i_{qs} \tag{7.9}$$

$$V_{qs} = R_s i_{qs} + L_q \frac{d}{dt} i_{qs} + \omega_s L_d i_{ds} + \omega_s \lambda_m \tag{7.10}$$

$$T_e = \frac{3}{2} * \frac{P}{2} \left[\lambda_m.i_{qs} + (L_d - L_q)i_{ds}.i_{qs} \right] \tag{7.11}$$

7.4.1 Pitch-angle controller

In the wind generating system, the rotor speed, torque, and the produced electrical energy are maintained at the desired levels by adjusting the pitch angle of wind turbine blades. It is possible to utilize the wind turbine at a constant speed by controlling the mechanical or electrical parts. Additionally, pitch control provides a safety mechanism as it limits the operating levels to the maximum capabilities of each machine. In the actual turbines, the power is controlled by pitch angle, rotor speed, and yaw angle. Although the air density affects the power production differently at all wind sites, it is interesting to mention that it decreases with increasing altitude in order to increase the produced power. A typical method for limiting the power under strong winds is to control the pitch angle [6].

In this project, a variable-pitch wind turbine is implemented. The turbine performance coefficient, Cp, is calculated by dividing the mechanical output power of the turbine by the wind speed, rotation speed, and pitch angle (beta). According to the data on wind turbines, Fig. 7.4 shows the output power of the test-case wind turbine following various wind speeds.

The developed pitch angle controller on the wind turbine is shown in Fig. 7.5. By using PI controllers, the mechanical rotation speed is measured and compared to the rated speed, and presented by Eq. (7.12), the proper angle is achieved.

$$\omega_{er} = \left(\omega_{pu} - \omega_m \right) \left(K_p + \frac{K_i}{s} \right) \tag{7.12}$$

7.4.2 Maximum power point tracking controller

The wind turbine is equipped with MPPT capability in order to enhance the efficiency of the wind generating system. The installed

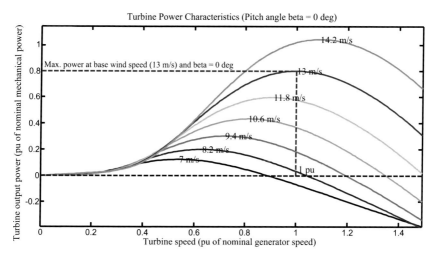

FIGURE 7.4   The output power of the test turbine at different wind speeds.

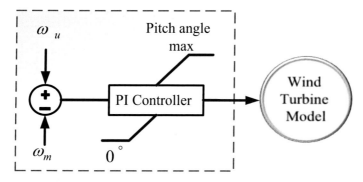

FIGURE 7.5   The designed pitch-angle controller.

MPPT harvests the maximum power from the wind-generating system. Therefore, Cp (β, λ) should keep its maximum level (i.e., Cpmax) under all possible wind speeds in the operating range; where the λ_{opt} is attained by keeping λ on the optimal level. The Cpmax achieves λ_{opt} by keeping the tip speed ratio (*TSR* λ) on the optimum level based on the relationship addressed by Eq. (7.13). As shown by Fig. 7.6, the three-phase power of the wind generating system is rectified by an uncontrolled three-phase rectifier. The MPPT unit controls the operation of the boost converter for extracting maximum power from the wind-generating system regardless of wind speed. Based on Eq. (7.14), the DC reference currents of the wind turbine are obtained by stabilizing the output DC voltage under different DC powers and angular speeds.

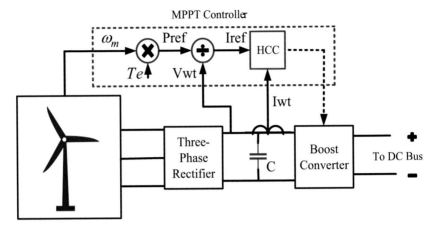

FIGURE 7.6 The designed maximum power point tracking control in a wind turbine.

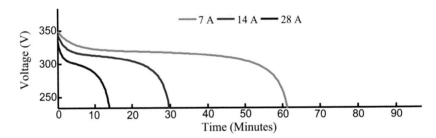

FIGURE 7.7 The discharge curve of the battery.

Then, the hysteresis current controller (HCC) unit compares the obtained value to the DC output current as expressed by Eq. (7.15) [10].

$$\omega_{mr} = \frac{\lambda_{opt}}{R} V \tag{7.13}$$

$$I_{WT}^{REF} = \frac{\omega_m \times Te}{V_{WT}} = \frac{P_{REF}}{V_{WT}} \tag{7.14}$$

$$I_{WT}^* = I_{WT}^{REF} - I_{WT} \tag{7.15}$$

7.5 Battery model

A dynamic model of a nickel–metal–hydride battery is used in the battery block. The parameters of the model can be changed depending on the discharge characteristics of the battery. Fig. 7.7 illustrates the typical discharge curve of the battery. The specifications of battery performance are extracted from MATLAB® software. In general, the battery voltage decreases in a shorter period of time when the current of the battery is greater than 7 A, which means the battery will reach full discharge sooner.

7.6 Coordinated control in island mode

Fig. 7.8 demonstrates the proposed strategy for coordinated control of parameters in the off-grid operation of the MG. This strategy allows the MG to supply the required active power by inverter. It is therefore possible for the control system to manage both the inverter and the DC/DC converters. In other words, the control system not only selects the charging or discharging state of the battery according to the required power but also delivers the DC power to the 380 V/50 Hz AC bus by triggering the three-phase inverter. During the power transmission, the control system sets the DC bus voltage on the reference value of 400 V so that the DC link voltage remains constant on the nominal value under battery charging–discharging operations.

It should be mentioned that DC link voltage stabilization is very important in off-grid operation. Because in this condition the changes in the load power can directly affect the voltage and make the DC voltage fluctuate. An effective control system should be able to stabilize the voltage and frequency in island mode. Since the MG under study has a DC structure, only the voltage stabilization is carried out here.

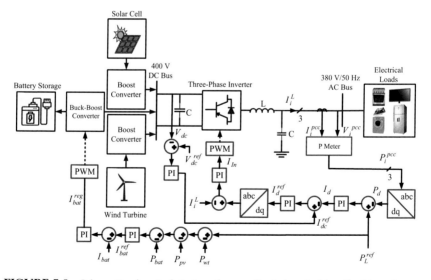

FIGURE 7.8 Schematic of control strategy for coordinated control in off-grid mode.

The relationship presented by Eq. (7.16) is used to determine the load power in the point of common coupling (PCC), which is then transferred to the d-axis by using Eq. (7.17). By comparing this power to the reference load power, the d-axis current is determined as Eq. (7.18). Based on this value, the d-axis reference current is calculated as Eq. (7.19). The DC reference current is expressed by Eq. (7.20) as well. The reference current on the d-axis is transferred to the ABC domain by using Eq. (7.21). The relationship presented in Eq. (7.22) can be used to determine the appropriate current for triggering the inverter, whereas the relationships given by Eqs. (7.23) and (7.24) are used to determine the appropriate current for triggering the buck-boost converter.

$$P_i^{pcc} = V_i^{pcc} \times I_i^{pcc} \times \cos\theta \qquad (7.16)$$
$$\forall i \in phase[a, b, c]$$

$$P_d = \frac{2}{3}\left(P_{pcca} \times \cos(\theta)\right) + \frac{2}{3}\left(P_{pccb} \times \cos\left(\theta - \frac{2\pi}{3}\right)\right) + $$
$$\frac{2}{3}\left(P_{pccc} \times \cos\left(\theta + \frac{2\pi}{3}\right)\right) \qquad (7.17)$$

$$I_d = \left(P_L^{ref} - P_d\right) \times \left(K_P + \frac{K_I}{s}\right) \qquad (7.18)$$

$$I_d^{ref} = \left(I_{dc}^{ref} - I_d\right) \times \left(K_P + \frac{K_I}{s}\right) \qquad (7.19)$$

$$I_{dc}^{ref} = \left(V_{dc}^{ref} - V_{dc}\right) \times \left(K_P + \frac{K_I}{s}\right) \qquad (7.20)$$

$$\begin{cases} I_a = (I_d^{ref} \times \cos(\theta)) \\ I_b = \left(I_d^{ref} \times \cos\left(\theta - \frac{2\pi}{3}\right)\right) \\ I_c = \left(I_d^{ref} \times \cos\left(\theta + \frac{2\pi}{3}\right)\right) \end{cases} \qquad (7.21)$$

$$I_{In} = \left(I_i - I_i^L\right) \times \left(K_P + \frac{K_I}{s}\right) \qquad (7.22)$$
$$\forall i \in phase[a, b, c]$$

$$I_{bat}^{ref} = \left(P_L^{ref} - P_{pv} - P_{wt} - P_{bat}\right) \times \left(K_P + \frac{K_I}{s}\right) \qquad (7.23)$$

$$I_{bat}^{reg} = \left(I_{bat}^{ref} - I_{bat}\right) \times \left(K_P + \frac{K_I}{s}\right) \qquad (7.24)$$

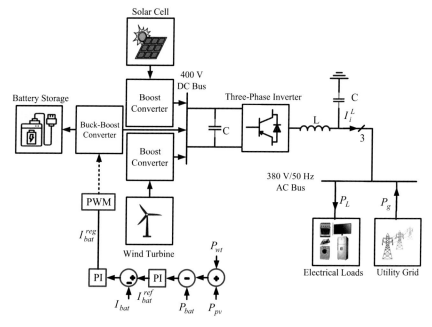

FIGURE 7.9 Schematic of control strategy for coordinated control in grid-tied.

7.7 Coordinated control in the grid-tied state

There is a more convenient control strategy when the loads are supplied by the network. In the grid-tied state, the MG loads are directly
supplied by the utility grid. Since the loads do not need the energy of
renewable resources on the DC bus, the control system is configured in
such a way that the power produced by wind and solar systems is
stored in the battery storage. For this purpose, according to Fig. 7.9, the
appropriate current for triggering the buck-boost converter is determined by Eqs. (7.25) and (7.26).

$$I_{bat}^{ref} = \left(P_{pv} + P_{wt} - P_{bat}\right) \times \left(K_P + \frac{K_I}{s}\right) \tag{7.25}$$

$$I_{bat}^{reg} = \left(I_{bat}^{ref} - I_{bat}\right) \times \left(K_P + \frac{K_I}{s}\right) \tag{7.26}$$

7.8 Numerical information and problem statement

Fig. 7.10 illustrates the proposed structure for the MG in detail. The
solar cell is linked to the DC bus by means of a boost converter which

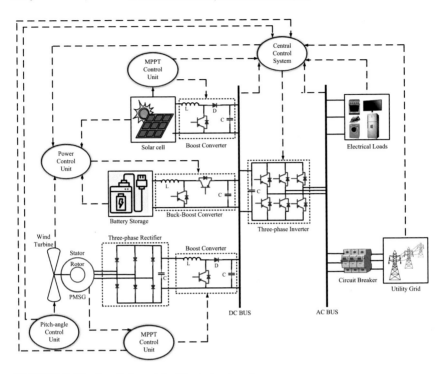

FIGURE 7.10 Detailed model of the test network.

operates based on the MPPT algorithm. The battery is as well linked to the DC bus by the use of a buck-boost converter. The three-phase rectifier converts the three-phase AC electricity generated by PMSG into proportional DC power in the wind-generating system. The wind turbine is linked to the DC link by the use of a boost converter equipped with an MPPT system. Moreover, the wind turbine is equipped with a mechanical pitch-angle controller. The DC and AC buses are linked by means of a three-phase inverter. The three-phase AC loads are modeled for the residential MG. Furthermore, the external network can be connected to the MG by the circuit breaker.

The central control system monitors all aspects of MG operation. The MPPT and pitch-angle units send their functional information to the central control system. By acquiring the information from the demand side and getting the status of off-grid or grid-tied operating mode, the central control system informs the power control unit about the power exchange between the DC and AC buses. In order to determine the operating mode of the battery, the power control unit makes the decision based on the received commands on the one hand and the level of produced power by resources on the other hand. Two working modes

TABLE 7.3 Electrical parameters of resources and battery in microgrid.

	Type	Voltage	Nominal power
PV	DC	160 (V)	10 (kW)
Wind turbine	DC	160 (V)	8.5 (kW)
Battery	DC	300 (V)	7 (Ampere-hours)

are defined for the battery including charging and discharging operating states. In the discharging mode, the DC bus receives the necessary power and the three-phase inverter converts it to the desired AC power with proper voltage and frequency. The central control system always monitors and regulates the DC and AC bus voltages. The DC bus voltage is 400 V and the AC bus voltage is 380 V/50 Hz. Table 7.3 shows the electrical parameters of the MG.

7.9 Simulations and numerical results

In order to evaluate the performance of the MG control system, The function of each controller is presented and discussed in detail.

7.9.1 MPPT operation in the wind and solar systems

In this scenario of operation, the solar irradiance is 620 w/m^2 from 0 to 0.5 seconds. The solar irradiance changes to 350 and 520 w/m^2 for 0.5–1.5 and 1.5–2 seconds, respectively. As shown in Fig. 7.11, the power extracted from the solar generating system varies per the changes in the solar irradiance. It is demonstrated when the MPPT control system is activated on the solar cell, more power can be obtained under different irradiations compared to the system without an MPPT control system. In other words, the MPPT unit results in a higher power extraction from PVs at a particular irradiation level. In the presence of MPPT, this power equals 6 kW in 0–0.5 seconds, 3.3 kW in 0.5–1.5 seconds, and 5.3 kW for 1.5–2 seconds, respectively. If MPPT is not activated, these powers reduce to 5.5, 2.7, and 4.8 kW, respectively. Besides, the changes in the output current of the solar unit can be seen in Fig. 7.11. As a result of using MPPT, the solar current decreases from 15 to 8 A and then increases to 12 A according to the variations in solar irradiation. However, without the MPPT unit, the solar current decreases from 13 to 4 A and afterward increases to 7 A.

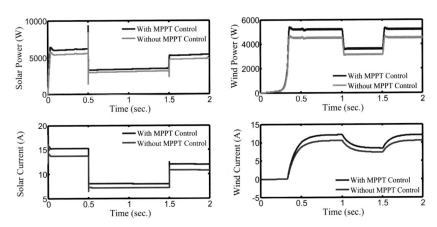

FIGURE 7.11 Power and current extracted from the wind and solar units with and without maximum power point tracking.

Fig. 7.11 also depicts the effects of MPPT on the power and current extracted from the wind turbine, where, the wind speed is equal to 11.5 m/s from 0.4 to 1 second, 8.5 m/s from 1 to 1.5 seconds, and finally, it increases to 11.5 m/s from 1.5 to 2 seconds. With the MPPT controller, the powers of the wind turbine from 0.4 to 1, 1 to 1.5, and 1.5 to 2 seconds are equal to 5.7, 3.8, and 5.7 kW, respectively. The powers reduce to 4.8, 3.4, and 4.8 kW, respectively, when MPPT is not utilized. Additionally, the variations caused by wind speed alterations are larger when MPPT is utilized compared to the system without MPPT.

7.9.2 Pitch angle controller performance

The variations in the wind speed will definitely change the harvested power from the wind turbine. In this case study, the rated wind speed is 12 m/s and the nominal power is extracted from the wind turbine when the wind speed is 12 m/s. For the lower speeds, less power is obtained. However, for the higher speeds that increase the power more than the rated power, it is necessary to limit the power to the rated level. The pitch angle control system is designed to deliver the rated power under wind speeds higher than the rated wind speed. This point protects the wind turbine from unauthorized speeds and possible damage to the mechanical and electrical parts.

Fig. 7.12 shows the function of the pitch angle controller. The wind speed from 0.8 to 1 second is equal to the rated wind speed of the turbine which is 12 m/s. After second 1, the wind speed increases from 12 to 15 m/s. From 1.5 to 1.8 seconds, the wind speed is 10 m/s. After second 1, the wind speed increases beyond the rated speed, and the pitch

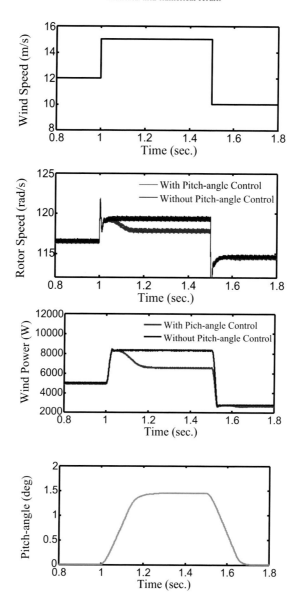

FIGURE 7.12   Power and rotor speed of wind turbine following wind speed alterations with and without pitch-angle controller.

angle controller takes action and limits the power extracted from the wind turbine by reducing the rotor speed and limiting power to 6.5 kW for 1 to 1.5 seconds. If this control system is not utilized, the extracted power and rotor speed increase to about 8 kW and 120 rad/s in this period of time. Such amplified power and rotor speed can cause serious damage to the mechanical parts as well as increase the current passing through the wiring and cause serious damage to the electrical insulations. It is clear that the pitch angle changes from 0 to 1.5 degrees in period when the wind speed is stronger than the nominal wind speed. From second 1.5 onward, because the wind speed decreases to 10 m/s, the extracted power also decreases and the pitch angle drives back to zero subsequently.

7.9.3 Control of off-grid microgrid with constant load/solar/ wind powers

In this case, the load power, wind speed, and solar irradiation are fixed and assumed as constant parameters. In this situation, the battery is charged or discharged based on the load power. The changes in the battery power are seen in Fig. 7.13. From 0 to 1 second, the load power is 5.8 kW and it must be supplied by wind and solar systems or by battery. From 0 to 0.5 second, the wind and solar powers are zero and the battery is producing 5.9 kW for supplying the loads. Some part of the produced power is wasted as losses. From 0.5 to 1 second, the solar cells produce 10 kW. As well, the wind turbine produces 5.7 kW from 0.7 to 1 second. Beyond second 0.5, the wind and solar systems are able to supply the loads, and the excess of their energy is stored by the battery. From 0.5 to 0.7 second, there is a 4.1 kW surplus of wind-solar energy and it is charged to the battery. From 0.7 to 1 second, the excess of wind-solar energy is 9.7 kW and it is properly stored by the battery. The powers of wind and solar units, the battery, and the energy loss are presented in Table 7.4.

Fig. 7.14 indicates the DC bus voltage and current. The voltage of the DC bus is fixed at 400 V in all time periods. The DC bus current is equal to 14 A. Fig. 7.15 shows the voltage and current of the three-phase load. Since the load is constant, these values are unvarying as well.

Figs. 7.16 and 7.17 show the voltage and current in wind and solar units, respectively. After the second 0.7, the voltage and current of the wind unit reach 400 V and 11 A, respectively. The voltage of the solar unit is equal to 400 V and its current is 25 A from 0.5 seconds onward.

7.9.4 Control of off-grid microgrid with variable load–wind powers

In this operating condition, according to Fig. 7.18, the load power and wind turbine power are changed and such alterations lead to changes in the battery power. The load power is equal to 5.8 kW from 0

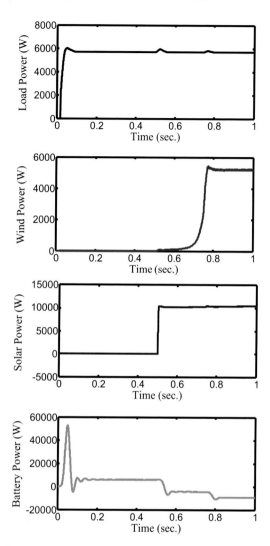

FIGURE 7.13   Powers under constant load power, wind speed, and solar irradiation.

to 1.5 seconds, and it is 8.5 kW from 1.5 to 2 seconds. The power of the wind turbine is 5.8, 3.9, and 2 kW for 0.4−1, 1−1.5, and 1.5−2 seconds, respectively. The solar irradiation is set at zero and the solar system has no production. The battery power is equal to 5.9 kW from 0 to 0.4 seconds. It is 0.2, 2.1, and 6.8 kW from 0.5−1, 1−1.5, and 1.5−2 seconds. The numerical details of the total consumptions and generations of electricity by wind−solar units, battery, and system losses are presented in Table 7.5.

TABLE 7.4   Power consumption−production under constant load−solar−wind powers.

Time (s)	Total consumption (kW)	Production of solar unit (kW)	Production of wind unit (kW)	Production of battery unit (kW)	Total losses (kW)
0−0.5	5.8	0	0	+5.9	0.1
0.5−0.7	5.8	10	0	−4.1	0.1
0.8−1	5.8	10	5.7	−9.7	0.2

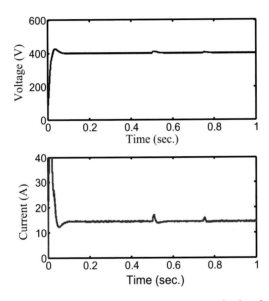

FIGURE 7.14   Voltage and current of DC bus under constant load−solar−wind powers.

Fig. 7.19 shows the DC bus voltage and current under the conditions of Fig. 7.18. The voltage of the bus is fixed at 400 V. The DC bus current is equal to 14 A from 0 to 1.5 seconds and it is 21.5 A from 1.5 to 2 seconds. Fig. 7.20 shows the voltage and current of the three-phase load. It is clear that with the increase of load after second 1.5, the load current increases, but the voltage produced by the three-phase inverter is still stable without fluctuations.

Fig. 7.21 shows the voltage and current of the wind unit. The voltage of the wind unit reaches 400 V at second 0.4. The current changes are proportional to the power from 0.5−1, 1−1.5, and 1.5−2 seconds, respectively.

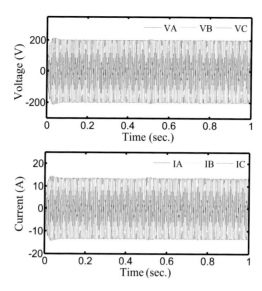

FIGURE 7.15   Voltage and current of load under constant load—solar—wind powers.

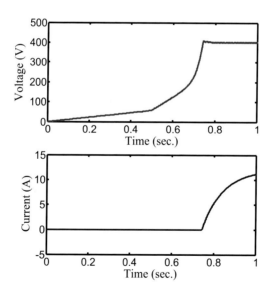

FIGURE 7.16   Voltage and current of wind turbine under constant load—solar—wind powers.

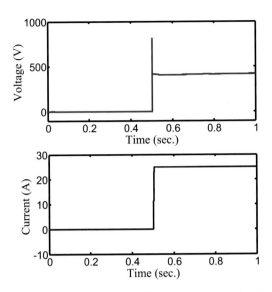

FIGURE 7.17 Voltage and current of solar cell under constant load−solar−wind powers.

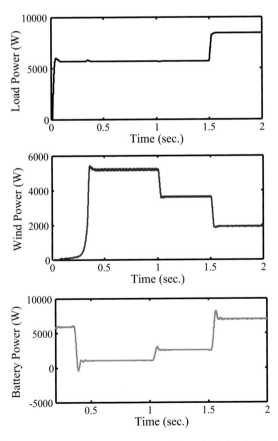

FIGURE 7.18 Load, wind and battery power under variable load−wind powers.

TABLE 7.5 Results under variable load-wind powers.

Time (s)	Total consumption (kW)	Production of solar unit (kW)	Production of wind unit (kW)	Production of battery unit (kW)	Total losses (kW)
0–0.4s	5.8	0	0.0	5.9	0.1
0.5–1	5.8	0	5.8	0.2	0.2
1–1.5	5.8	0	3.9	2.1	0.2
1.5–2	8.5	0	2.0	6.8	0.3

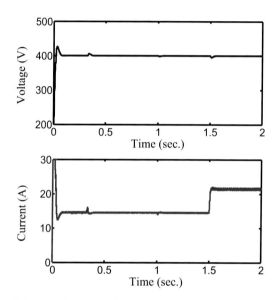

FIGURE 7.19 Voltage and current of DC bus under variable load–wind powers.

7.9.5 Control of off-grid microgrid with variable load–solar powers

In this operating condition, according to Fig. 7.22, the wind speed is assumed to be zero and the output power of the wind generating system is set at zero. The load power from 0 to 1 second is equal to 5.5 kW, and it is 11 kW from 1 to 2 seconds. The sunlight is applied from seconds 0.5. The irradiation decreases at second 1 and increases at second 1.5. Therefore, according to the changes in solar irradiation, the power extracted from the solar cell is different. The power is zero from 0 to 0.5 seconds, it becomes 3 kW from 0.5 to 1 second, and it is equal to 0.3 and 5.5 kW in the next steps, respectively. According to the changes in the produced power of solar cells and the power consumed by the load,

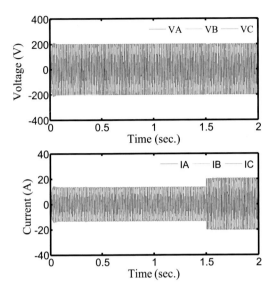

FIGURE 7.20   Voltage and current of load under variable load–wind powers.

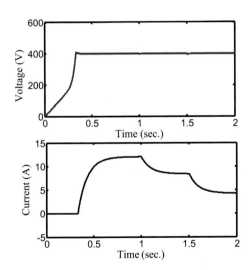

FIGURE 7.21   Voltage and current of wind turbine under variable load–wind powers.

the mismatch between them is addressed by the battery. For this purpose, the battery power from 0 to 0.5 seconds is 5.6 kW, from 0.5 to 1 second it is 2.7 kW, and from 1 to 1.5 and 1.5 to 2 seconds it is 10.9 and 5.9 kW, respectively. Table 7.6 shows the powers of load, solar cell, and battery alongside the total energy loss.

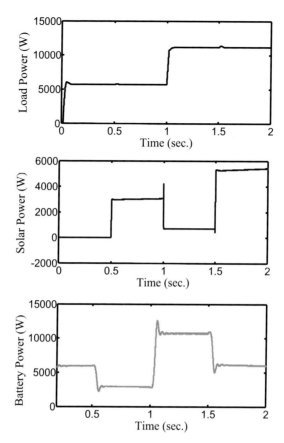

FIGURE 7.22   Load, solar, and battery power following variable load-solar powers.

TABLE 7.6   Results under variable load-solar powers.

Time (s)	Total consumption (kW)	Production of solar unit (kW)	Production of wind unit (kW)	Production of battery unit (kW)	Total losses (kW)
0–0.5	5.5	0	0	5.6	0.1
0.5–1	5.5	3	0	2.7	0.2
1–1.5	11	0.3	0	10.9	0.2
1.5–2	11	5.5	0	5.7	0.2

Fig. 7.23 shows the voltage and current produced by the solar unit under different solar irradiations. With the start of solar irradiation from second 0.5, the voltage of the boost converter on the solar cells is

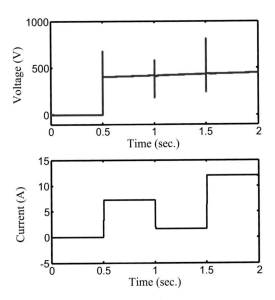

FIGURE 7.23   Voltage and current of solar cell under variable load−solar powers.

fixed at 400 V. The voltage remains constant under all changes in the power production in the next moments. On the other hand, the output current of the converter changes proportional to the power at seconds 0.5, 1, and 1.5.

7.9.6 Control of off-grid microgrid with variable wind−solar powers

Under this operating condition, the load power remains constant. The power extracted from the wind turbine changes in two steps and the power obtained from the solar cell changes in three steps. Fig. 7.24 shows that the battery power changes are proportional to the changes in the solar−wind powers at different times. The numerical details of powers are presented in Table 7.7.

Figs. 7.25 and 7.26 show the voltages and currents of solar cell and wind turbine units, respectively. The currents change according to the power alterations, but the voltages are fixed at 400 V.

7.9.7 Control of off-grid microgrid with variable load−solar−wind powers

In this section, the load, solar and wind powers change in several steps. The last step of changes shows a condition in which the load

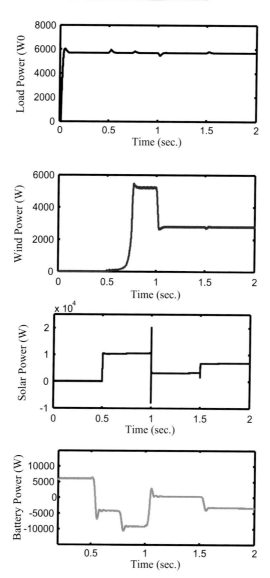

FIGURE 7.24   Load, wind, solar cell, and battery power under variable wind—solar powers.

power increases but both the solar and wind powers decrease to zero. This operating condition is developed in order to demonstrate the capability of the control system for supplying the load when wind and solar energy are unavailable. In such a situation, the load is solely supplied by the battery. Fig. 7.27 shows the battery operation under variable load—solar—wind powers and the numerical data are presented in

TABLE 7.7 Results under variable wind-solar powers.

Time (s)	Total consumption (kW)	Production of solar unit (kW)	Production of wind unit (kW)	Production of battery unit (kW)	Total losses (kW)
0–0.5	5.9	0	0	6.0	0.1
0.5–0.7	5.9	10	0	−4.5	0.3
0.8–1	5.9	10	5.5	−9.4	0.2
1–1.5	5.9	2.5	2.8	0.8	0.2
1.5–2	5.9	7.4	2.8	−4.1	0.2

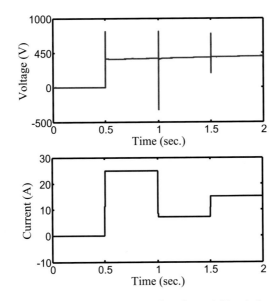

FIGURE 7.25   Voltage and current of solar cell under variable wind–solar powers.

Table 7.8. The solar and wind powers became zero at second 1.5 and the battery rapidly responds in order to handle the mismatch between production and consumption.

Fig. 7.28 shows the DC bus voltage and current. Following all changes in the load and the power generation, the bus voltage is still regulated at 400 V, but the current changes in accordance with the power alterations.

Fig. 7.29 shows the voltage and current of the three-phase load. The current changes according to the load variations at seconds 1 and 1.5, but the voltage, which is supplied by the three-phase inverter, remains constant and stable. Therefore, the load can continue operation without any issues in the power quality.

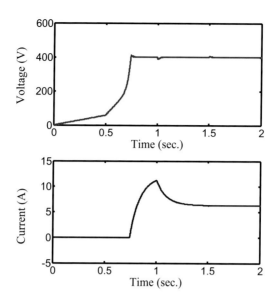

FIGURE 7.26   Voltage and current of wind turbine under variable wind−solar powers.

7.9.8 Control of grid-connected microgrid

The previous scenarios investigated and analyzed the performance of the MG in the off-grid operation. In this section, the grid-connected operation is controlled and evaluated. In the grid-connected, the produced powers by solar and wind units are considered according to the previous conditions. The load increases in three steps and the MG is connected to the network at second 1. The results are depicted in Fig. 7.30. From second 1 onward, the external network is responsible for supplying the load, and the output power of solar and wind units is stored in the battery through the charging process. More details are provided in Table 7.9.

Fig. 7.31 shows the DC bus voltage and current. The bus voltage is fixed at 400 V. From second 1 onward since the network is responsible for supplying the load, the DC bus current becomes zero. In other words, no power is transferred from the DC bus side to the AC bus side to supply the load but the load on the AC bus is supplied from the grid side.

7.9.9 Connecting microgrid to the grid for mutual power exchange

In the previous sections, the performance of the MG was studied in both the island and grid-connected modes. The designed control system

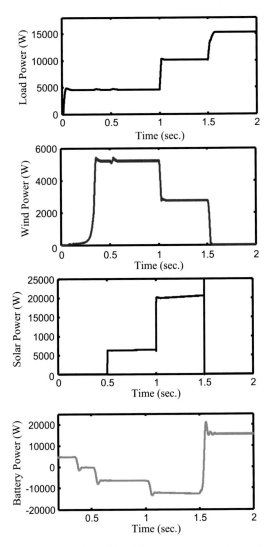

FIGURE 7.27 Load, wind, solar cell and battery power under variable load–solar–wind powers.

properly controlled the resources and battery under various operating conditions. In the off-grid mode, the control system stabilized the DC link voltage on the nominal value under all operating conditions and loadings. The DC link responded to the changes in the load, wind, or solar power via the three-phase inverter. When the MG was connected to the external grid, the loads were supplied by the grid side, and the solar–wind power was stored in the battery.

TABLE 7.8 Results under variable load—solar—wind powers.

Time (s)	Total consumption (kW)	Production of solar unit (kW)	Production of wind unit (kW)	Production of battery unit (kW)	Total losses (kW)
0–0.3	5.9	0	0	6.1	0.2
0.3–0.5	5.9	0	5.5	0.5	0.1
0.5–1	5.9	6	5.5	−5.4	0.2
1–1.5	10	20	2.7	−12.5	0.2
1.5–2	15	0	0	15.3	0.3

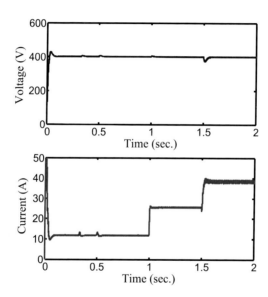

FIGURE 7.28 Voltage and current of DC bus under variable load—solar—wind powers.

The aforementioned strategy is one of the simplest methods for connecting the MG to the external network, where the power is only taken from the grid. The MG can be connected to the upstream grid in such a way that it is able to inject power into the external grid. Therefore, one of the objectives of connecting the MG to the external grid can be participating in the electricity market. By participating in the market paradigm, the MG can trade electricity with the upstream network according to the electricity market contracts. In this case, the MG not only assists the grid but also can make a profit by trading electricity with the grid. The MG is also able to participate in the ancillary services

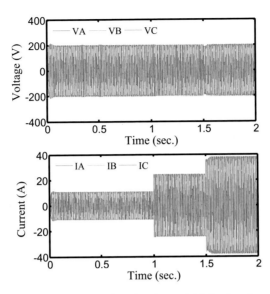

FIGURE 7.29 Voltage and current of load under variable load—solar—wind powers.

of the electricity market such as frequency regulation, reactive power compensation, and voltage regulation. All of those objectives need to be implemented through accurate control loops. Each objective needs to be realized by a separate control system.

In this section, the connection of the MG to the network is evaluated with the aim of injecting flexible and controllable active power into the network. For this purpose, the schematic shown in Fig. 7.32 is developed. Two three-phase inverters are used to connect the DC and AC buses to each other. Inverter 1 is responsible for exchanging and injecting power into the network. In order to achieve this purpose, it has a separate control system that operates under the supervision of central control system. The three-phase inverter 2 is also implemented with the purpose of supplying three-phase loads on the AC bus. With this structure, not only the loads are supplied but also the MG is able to inject scheduled active power into the upstream network.

It is assumed that the MG has to send a fixed power to the external grid based on the contract with the grid operator. As a result, the control system in the MG should control and regulate the power between the MG and the external grid on the predetermined level under all load or renewable energy alterations in the MG. The MG is controlled to send the predetermined active power to the external grid under all operating conditions. This predetermined power may be changed by the network operator and it would be set on a new power level. The MG must have the capability to alter the exchanged power based on the

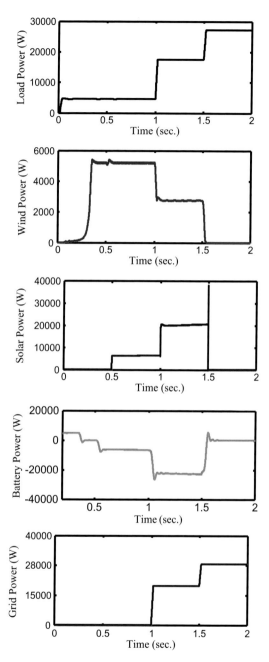

FIGURE 7.30 Grid, load, wind, solar cell, and battery power in the grid-connected state.

TABLE 7.9 Numerical values of powers in the grid-connected state.

Time (s)	Total consumption (kW)	Production of solar unit (kW)	Production of wind unit (kW)	Production of battery unit (kW)	Power from grid (kW)	Total losses (kW)
0–0.3	6	0	0.0	6.10	0	0.1
0.3–0.5	6	0	5.5	0.60	0	0.1
0.5–1	6	6	5.5	−5.30	0	0.2
1–1.5	18	20	2.7	−22.5	18	0.2
1.5–2	28	0	0.0	0.00	28	0.0

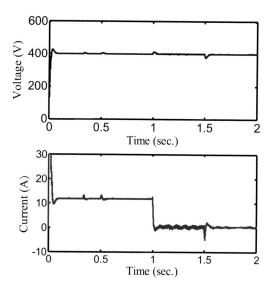

FIGURE 7.31   Voltage and current of DC bus in grid-connected state.

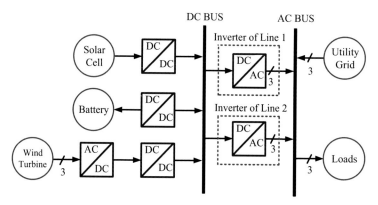

FIGURE 7.32   New structure for connecting microgrid to an external grid.

command of the network operator. As a result, the flexible active power is delivered to the upstream network by MG through inverter 1. The MG is controlled by the central control system. The reference active power is continuously changed by the central control system based on the conditions of load power in order to exchange electricity with the network according to the contract. The developed strategy is defined as follows; if the load current is less than 10 A, the injected power to the network is 30 kW, if the load current is more than 10 A, the injected power is reduced to 20 kW and finally if the load current increases to more than 20 A, the MG delivers only 10 kW to the network. In other words, the MG offers a flexible capacity for exchanging with the network according to the load power in the MG. When the MG needs to supply large local loads, the traded power with the network is reduced and the power is devoted to the local loads.

In Fig. 7.33, the DC link is connected to inverter 1. Inverter 1 is connected to the network through the LC filter, three-phase transformer, and line. Index 1 in Fig. 7.33 refers to the control system of inverter 1. As expressed by Eq. (7.27), the three-phase inverter 1 provides a share of active power. The three-phase power of the MG is transferred from the ABC domain to the $d-q$ space by Eq. (7.28). The DC control unit obtains the DC reference current by Eq. (7.29). The d-axis reference current is calculated by Eq. (7.30) and the final d-axis current is obtained as specified by Eq. (7.31). It is transferred back to the ABC domain by Eq. (7.32). By comparing these currents to the three-phase currents of the line based on Eq. (7.33), the required current of inverter 1 is calculated, which provides proper PWM pulses for inverter 1.

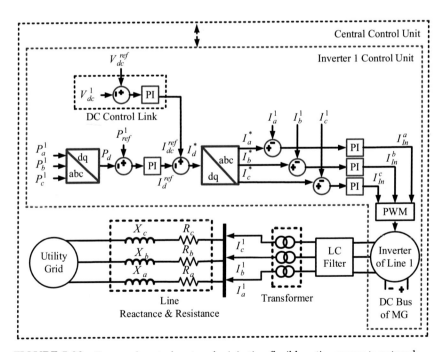

FIGURE 7.33 Proposed control system for injecting flexible active power to network.

$$P_i = V_i \times I_i \times \cos\theta_i$$
$$\forall i \in Phase[a, b, c] \tag{7.27}$$

$$P_d = \frac{2}{3}\left[P_i \times \cos\theta + P_i \times \cos\left(\theta - \frac{2\pi}{3}\right) + P_i \times \cos\left(\theta + \frac{2\pi}{3}\right)\right] \tag{7.28}$$
$$\forall i \in Phase[a, b, c]$$

$$I_d^{ref} = K_P(P_{ref} - P_d) + K_I \int (P_{ref} - P_d)dt \tag{7.29}$$

$$I_{dc}^{ref} = K_P(V_{dc}^{ref} - V_{dc}) + K_I \int (V_{dc}^{ref} - V_{dc})dt \tag{7.30}$$

$$I_d^* = I_{dc}^{ref} - I_d^{ref} \tag{7.31}$$

$$\begin{cases} I_i^* = [I_d^* \times \cos\theta] \\ I_i^* = \left[I_d^* \times \cos\left(\theta - \frac{2\pi}{3}\right)\right] \\ I_i^* = \left[I_d^* \times \cos\left(\theta + \frac{2\pi}{3}\right)\right] \end{cases} \forall i \in Phase[a, b, c] \tag{7.32}$$

$$I_{In}^i = K_P(I_i^* - I_i) + K_I \int \left(I_i^* - I_i^j\right)dt \tag{7.33}$$
$$\forall i \in Phase[a, b, c]$$

The operating conditions are considered in such a manner that the MG load changes at different moments in order to demonstrate the ability of the proposed control system to inject flexible power into the upstream network by using inverter 1. Inverter 2 supplies the load power, while inverter 1 injects flexible active power to the network according to the amount of load supplied by inverter 2. Fig. 7.34 shows the load current supplied by inverter 2. The load current is 15 A from 0.2 to 0.4 seconds

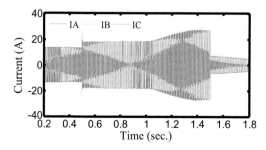

FIGURE 7.34 Load current in the microgrid supplied by the second inverter.

and it is 18 A from 0.5 to 1 second. From 1 to 1.5 seconds, the load current increases in the form of a slope and reaches 21 A at second 1.1. From second 1.5 onward, the load current decreases to 5 A.

Fig. 7.35 shows the changes in power injected into the network. It should be noted that inverter 1 is connected to the grid at second 0.5 and the proposed strategy is operated after second 0.5. The beginning time is arbitrary and can be set at any time based on the operator's request. According to Fig. 7.35, after beginning the process of injecting power to the grid through inverter 1, the injected power is equal to 20 kW until second 1 because the load current at this time period is greater than 10 A and less than 20 A. According to the designed plan, the aforementioned level of load current results in 20 kW power injection to the network.

Together with increasing the load current after second 1, the power injected to the network decreases until second 1.1, because the load current is more than 20 A and according to the proposed strategy, the MG reduces the traded power with the network to 10 kW. This condition remains constant until the second 1.5 and at this moment, since the load current is less than 10 A, the central control system increases the injected power to 30 kW.

Based on this strategy, the control system changes the injected power to the network according to the load powers. When the load demand in the MG is low, the MG uses this opportunity to sell more power to the grid. Once the local loads increase, the MG reduces the traded power with the grid and supplies the local loads. Since loads of the MG change continuously, the designed control system must uninterruptedly monitor the load power and regulate the exchanged power with the grid consequently.

Fig. 7.36 shows the alterations in the current delivered to the network under the flexible power injection presented in Fig. 7.35. This current shows similar changes and it is altered proportionally to the power variations.

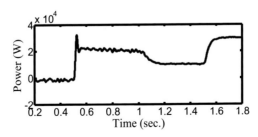

FIGURE 7.35   Flexible power injected into the network by microgrid through inverter 1.

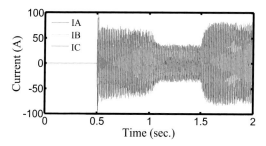

FIGURE 7.36 Changes in the injected current to the network.

7.10 Advanced models for microgrid control

Besides the introduced models for MG control, there is a broad scope of research studies on the design and implementation of control systems in MGs. Some of those methods are summarized here.

The load and energy regulation in the residential DC MG integrated with fuel cells, solar cells, and batteries is presented by [1]. The DC, single-phase, and three-phase AC loads are integrated at 50 Hz frequency. The MG is linked to the 60 Hz three-phase AC utility grid and it is able to send 8 kW constant power to the external network. A well-organized multibus topology is used for the MG and it is created by different AC/DC buses for supplying the loads and operating the resources.

The model developed by [2] provides a new concept for the coordination of central battery storage with decentralized hybrid renewable systems in different buildings. The developed system consists of 5 buildings that are equipped with a central 300 V, 6.5 Ampere-hours battery, and several DG resources. The central battery can be linked to any of the buildings for exchanging power. In order to increase the resilience of the system when the battery operates with powers above 50 kW, the designed control system is permitted to connect the buildings to the external AC network. The desired network performance is investigated under different scenarios in different working conditions.

An energy management model in a zero-energy building equipped with renewable generating systems, ES devices, AC–DC energy consumers, and critical/noncritical loads is designed by [3]. The developed model controls the performance of solar cells and fuel cells, schedules the battery charge–discharge, adjusts the load energy, and regulates the bus voltage. The faults, events, and cyber-attacks are distinguished and correctly handled. The AC loads with dissimilar voltage and frequency levels are operated. The critical and noncritical loads are modeled and the energy management system successfully supplies the critical loads following any situation. The fault detection systems are applied on both the AC and DC buses for monitoring the AC–DC voltages in healthy and faulty operating conditions. The outage detection unit is applied for compensating the shortage of energy caused by unavailable resources. The cyber-attack detection system is applied to reveal the attacks on the parameters including solar cell power, fuel cell power, AC voltage, and DC voltage.

The research presented by [4] develops a new control structure on an AC–DC ring-bus off-grid MG under unbalanced and nonlinear operating conditions. The DC bus is equipped with one fuel cell, one wind turbine, and two batteries. The DC bus is arranged to be separated into two subsections, the first section is equipped with the fuel cell and one battery, and the second section is equipped with the wind turbine and the other battery. The AC bus is as well arranged to be separated into two sections, where, the diesel generator and the unbalanced load are located on the first section and the nonlinear load together with the wind turbine are equipped on the other section. The control system is able to balance the unbalanced loads. As well, the control system deals with the harmonics of nonlinear loads and enhances the voltage profile.

Innovative control schemes are presented for balancing the time-varying and unbalanced loads using the fuel cells and the battery [5]. In grid-connected operation, the fuel cell and battery are optimally operated to balance the unbalanced loads, where the three-phase loads with both active and reactive powers are modeled and balanced. In the island operating condition, the diesel generator is optimally utilized to supply the active and reactive powers of the three-phase load, and the imbalances are addressed by the optimal operation of the fuel cell and the battery. The voltage magnitude is as well enhanced by adding sufficient reactive power to the system by means of local resources.

A flexible control system is designed by [6] in the islanded MG with AC and DC buses. The DC bus is equipped with a fuel cell, solar generating system, and battery storage system, and the AC bus is supported by a diesel generator and a wind-generating system. The AC bus is designed with three subbuses, which are allowed to continue their operation when one of the subbuses is faced with an outage. The link between DC and AC buses is formed by using two parallel three-phase lines, where each line is formed by three single-phase inverters.

The model addressed by [10] presents a comprehensive control scheme for the multi-MG system. This system is designed by several sub-MGs. In each sub-MG, the DC and AC buses are linked to each other through three single-phase DC-to-AC converters. The hybrid wind–solar generating system equipped with MPPT is applied in the MGs. The MGs are intended to work in both the off-grid and grid-tied operating states. In the islanded conditions, the subgrids support each other for enhancing the resilience of the system. Such a model increases the capability of load recovery after faults on the DC or AC sections.

7.11 Conclusions

This chapter investigated and analyzed the design and implementation of control systems in residential MGs for managing batteries and resources. The developed control system controlled the power of all resources including battery, local wind turbine, local solar cell, utility grid, and loads. Many operating strategies were expressed and implemented in the projected control system. The control system successfully utilized the defined strategies for managing power in the MG under different operating conditions such as healthy, faulty, outage, uncertainty, and alterations in the parameters.

Furthermore, wind and solar generating systems were equipped with an MPPT system. Simulation of various operating scenarios demonstrated that the designed control system is able to supply the loads under various loading-generating conditions as well as different changes in the off-grid and grid-connected operating modes. Moreover, some advanced control models and structures which are used in the MGs were reviewed and discussed.

References

[1] Faraji H, Nosratabadi SM, Hemmati R. AC unbalanced and DC load management in multi-bus residential microgrid integrated with hybrid capacity resources. Energy 2022;252:124070.

[2] Faraji H, Hemmati R. A novel resilient concept for district energy system based on central battery and decentral hybrid generating resources. International Journal of Energy Research 2022;46.

[3] Hemmati R, Faraji H. Identification of cyber-attack/outage/fault in zero-energy building with load and energy management strategies. Journal of Energy Storage 2022;50:104290.

[4] Hemmati R, Faraji H. Energy management in standalone AC/DC microgrid with sectionalized ring bus and hybrid resources. International Journal of Energy Research 2022.

[5] Hemmati R, Faraji H. Single-phase control of three-phase fuelcell-battery under unbalanced conditions considering off-grid and grid-tied states. Electric Power Systems Research 2021;194:107112.

[6] Hemmati R, Faraji H. Resilient control of hybrid microgrid with harmonic-unbalanced loads under outages and faults. IET Renewable Power Generation 2022;16:565−80.

[7] Salman S, Ai X, Wu Z. Design of a P-&-O algorithm based MPPT charge controller for a stand-alone 200W PV system. Protection and Control of Modern Power Systems 2018;3.

[8] Chen J, Yao W, Zhang C-K, Ren Y, Jiang L. Design of robust MPPT controller for grid-connected PMSG-based wind turbine via perturbation observation based nonlinear adaptive control. Renewable Energy 2019;134:478−95.

[9] Zouheyr D, Lotfi B, Abdelmadjid B. Improved hardware implementation of a TSR based MPPT algorithm for a low cost connected wind turbine emulator under unbalanced wind speeds. Energy 2021;232:121039.

[10] Hemmati R, Faraji H, Beigvand NY. Multilevel and advanced control scheme for multimicrogrid under healthy-faulty and islanded-connected conditions. IEEE Systems Journal 2021;1−9.

[11] Bhukya R, Shanmugasundaram N. Performance investigation on novel MPPT controller in solar photovoltaic system. Materials Today: Proceedings 2021.

[12] Wang J, Bo D, Ma X, Zhang Y, Li Z, Miao Q. Adaptive back-stepping control for a permanent magnet synchronous generator wind energy conversion system. International Journal of Hydrogen Energy 2019;44:3240−9.

Index

Note: Page numbers followed by "*f*" and "*t*" refer to figures and tables, respectively.